Springer Texts in Statistics

Advisors:
George Casella   Stephen Fienberg   Ingram Olkin

Springer Texts in Statistics

(continued after index)

Brian S. Everitt

An R and S-PLUS®
Companion to Multivariate
Analysis

With 59 Figures

 Springer

Brian Sidney Everitt, BSc, MSc
Emeritus Professor, King's College, London, UK

British Library Cataloguing in Publication Data
Everitt, Brian
   An R and S-PLUS® companion to multivariate analysis.
   (Springer texts in statistics)
   1. S-PLUS (Computer file) 2. Multivariate analysis-Computer programs. 3. Multivariate analysis-Data
   processing
   I. Title
   519.5′35′0285

   ISBN 1852338822

Library of Congress Cataloging-in-Publication Data
Everitt, Brian.
   An R and S-PLUS® companion to multivariate analysis/Brian S. Everitt.
      p. cm.—(Springer texts in statistics)
   Includes bibliographical references and index.
   ISBN 1-85233-882-2 (alk. paper)
      1. Multivariate analysis. 2. S-Plus. 3. R (Computer program language) I. Title. II. Series.

QA278.E926 2004
519.5′35—dc22   2004054963

ISBN 1-85233-882-2
Springer Science+Business Media
springeronline.com

© Springer-Verlag London Limited 2005

Printed in the United States of America
Typeset by Techset Composition Limited
12/3830-543210 Printed on acid-free paper   SPIN 10969380

To my dear daughters, Joanna and Rachel

Preface

The majority of data sets collected by researchers in all disciplines are multivariate. In a few cases it may be sensible to isolate each variable and study it separately, but in most cases all the variables need to be examined simultaneously in order to fully grasp the structure and key features of the data. For this purpose, one or another method of multivariate analysis might be most helpful, and it is with such methods that this book is largely concerned.

Multivariate analysis includes methods both for describing and exploring such data and for making formal inferences about them. The aim of all the techniques is, in a general sense, to display or extract the signal in the data in the presence of noise, and to find out what the data show us in the midst of their apparent chaos.

The computations involved in applying most multivariate techniques are considerable, and their routine use requires a suitable software package. In addition, most analyses of multivariate data should involve the construction of appropriate graphs and diagrams and this will also need to be carried out by the same package. R and S-PLUS® are statistical computing environments, incorporating implementations of the S programming language. Both are powerful, flexible, and, in addition, have excellent graphical facilities. It is for these reasons that they appear in this book. R is available free through the Internet under the General Public License; see R Development Core Team (2004), R: A Language and Environment for Statistical Computing, R Foundation for Statistical Computing, Vienna, Austria, or visit their website www.R-project.org. S-PLUS is a registered trademark of Insightful Corporation, www.insightful.com. It is distributed in the United Kingdom by

Insightful Limited
5th Floor
Network House
Basing View
Basingstoke
Hampshire
RG21 4HG

Tel: +44 (0) 1256 339800

Fax: +44 (0) 1256 339839
info.uk@insightful.com

and in the United States by

Insightful Corporation
1700 Westlake Avenue North
Suite 500
Seattle, WA 98109-3044

Tel: (206) 283-8802
Fax: (206) 283-8691
info@insightful.com

We assume that readers have had some experience using either R or S-PLUS, although they are not assumed to be experts. If, however, they require to learn more about either program, we recommend Dalgaard (2002) for R and Krause and Olson (2002) for S-PLUS. An appendix very briefly describes some of the main features of the packages, but is intended primarily as nothing more than an *aide memoire*. One of the most powerful features of both R and S-PLUS (particularly the former) is the increasing number of functions being written and made available by the user community. In R, for example, CRAN (Comprehensive R Archive Network) collects libraries of functions for a vast variety of applications. Details of the libraries that can be used within R can be found by typing in `help.start()`. Additional libraries can be accessed by clicking on **Packages** followed by **Load package** and then selecting from the list presented.

In this book we concentrate on what might be termed the "core" multivariate methodology, although mention will be made of recent developments where these are considered relevant and useful. Some basic theory is given for each technique described but not the complete theoretical details; this theory is separated out into "displays." Suitable R and S-PLUS code (which is often identical) is given for each application. All data sets and code used in the book can be found at http://biostatistics.iop.kcl.ac.uk/publications/everitt/. In addition, this site contains the code for a number of functions written by the author and used at a number of places in the book. These can no doubt be greatly improved! After the data files have been downloaded by the reader, they can be read using the `source` function

R: `name<-source("path")$value`

For example,

`huswif<-source("c:\\allwork\\rsplus\\chap1huswif.dat")$value`

S-PLUS: `name<-source("path")`

For example,

`huswif<-source("c:\\allwork\\rsplus\\chap1huswif.dat")`

Since the output from S-PLUS and R is not their most compelling or attractive feature, such output has often been edited in the text and the results then displayed in a different form from this output to make them more readable; on a few occasions, however, the exact output itself is given. In one or two places the "click-and-point" features of the S-PLUS GUI are illustrated.

This book is aimed at students in applied statistics courses at both the undergraduate and postgraduate levels. It is also hoped that many applied statisticians dealing with multivariate data will find something of interest.

Since this book contains the word "companion" in the title, prospective readers may legitimately ask "companion to what?" The answer is, to a multivariate analysis textbook that covers the theory of each method in more detail but does not incorporate the use of any specific software. Some examples are Mardia, Kent, and Bibby (1979), Everitt and Dunn (2002), and Johnson and Wichern (2003).

I am very grateful to Dr. Torsten Hothorn for his advice about using R and for pointing out errors in my initial code. Any errors that remain, of course, are entirely due to me.

Finally I would like to thank my secretary, Harriet Meteyard, who, as always, provided both expertise and support during the writing of this book.

London, UK   Brian S. Everitt

Contents

1
Multivariate Data and Multivariate Analysis

1.1 Introduction

Multivariate data arise when researchers measure several variables on each "unit" in their sample. The majority of data sets collected by researchers in all disciplines are multivariate. Although in some cases it may make sense to isolate each variable and study it separately, in the main it does not. In most instances the variables are related in such a way that when analyzed in isolation they may often fail to reveal the full structure of the data. With the great majority of multivariate data sets, *all* the variables need to be examined simultaneously in order to uncover the patterns and key features in the data. Hence the need for the collection of multivariate analysis techniques with which this book is concerned.

Multivariate analysis includes methods that are largely descriptive and others that are primarily inferential. The aim of all the procedures, in a very general sense, is to display or extract any "signal" in the data in the presence of noise, and to discover what the data has to tell us.

1.2 Types of Data

Most multivariate data sets have a common form, and consist of a data matrix, the rows of which contain the units in the sample, and the columns of which refer to the variables measured on each unit. Symbolically a set of multivariate data can be represented by the matrix, $\mathbf{X}$, given by

$$\mathbf{X} = \begin{bmatrix} x_{11} & x_{12} & \cdots & x_{1q} \\ \vdots & & \ddots & \\ x_{n1} & x_{n2} & \cdots & x_{nq} \end{bmatrix}$$

where n is the number of units in the sample, q is the number of variables measured on each unit, and x_{ij} denotes the value of the jth variable for the ith unit.

The units in a multivariate data set will often be individual people, for example, patients in a medical investigation, or subjects in a market research study. But they can also be skulls, pottery, countries, products, to name only four possibilities. In all cases the units are often referred to simply as "individuals," a term we shall generally adopt in this book.

A hypothetical example of a multivariate data matrix is given in Table 1.1. Here $n = 10$, $q = 7$, and, for example, $x_{33} = 135$. These data illustrate that the variables that make up a set of multivariate data will not necessarily all be of the same type. Four levels of measurement are often distinguished;

- *Nominal*—Unordered categorical variables. Examples include treatment allocation, the sex of the respondent, hair color, presence or absence of depression, and so on.
- *Ordinal*—Where there is an ordering but no implication of equal distance between the different points of the scale. Examples include social class and self-perception of health (each coded from I to V, say), and educational level (e.g., no schooling, primary, secondary, or tertiary education).
- *Interval*—Where there are equal differences between successive points on the scale, but the position of zero is arbitrary. The classic example is the measurement of temperature using the Celsius or Fahrenheit scales.
- *Ratio*—The highest level of measurement, where one can investigate the *relative magnitude* of scores as well as the differences between them. The position of zero is fixed. The classic example is the absolute measure of temperature (in Kelvin, for example) but other common examples include age (or any other time from a fixed event), weight and length.

The qualitative information in Table 1.1 could have been presented in terms of numerical codes (as often would be the case in a multivariate data set) such that sex = 1 for males and sex = 2 for females, for example, or health = 5 when very good and health = 1 for very poor, and so on. But it is vital that both the user and consumer of these data appreciate that the same numerical codes (1, say) will often convey completely different information.

In many statistical textbooks discussion of different types of measurements is often followed by recommendations as to which statistical techniques are suitable

Table 1.1 Hypothetical Set of Multivariate Data

Individual	Sex	Age (yr)	IQ	Depression	Health	Weight (lb)
1	Male	21	120	Yes	Very good	150
2	Male	43	NK	No	Very good	160
3	Male	22	135	No	Average	135
4	Male	86	150	No	Very poor	140
5	Male	60	92	Yes	Good	110
6	Female	16	130	Yes	Good	110
7	Female	NK	150	Yes	Very good	120
8	Female	43	NK	Yes	Average	120
9	Female	22	84	No	Average	105
10	Female	80	70	No	Good	100

NOTE: NK = not known.

for each type; for example, analyses of nominal data should be limited to summary statistics such as the number of cases, the mode, and so on. And in the analysis of ordinal data, means and standard deviations are not really suitable. But Velleman and Wilkinson (1993) make the important point that restricting the choice of statistical methods in this way may be a dangerous practice for data analysis; the measurement taxonomy described is often too strict to apply to real-world data. This is not the place for a detailed discussion of measurement, but we take a fairly pragmatic approach to such problems. For example, we will often not agonize over treating variables such as a measure of depression, anxiety, or intelligence as if they were interval-scaled, although strictly they fit into the ordinal category described above.

Table 1.1 also illustrates one of the problems often faced by statisticians undertaking statistical analysis in general, and multivariate analysis in particular, namely the presence of *missing values* in the data, that is, observations and measurements that should have been recorded, but, for one reason or another, were not. Often when faced with missing values, practitioners simply resort to analyzing only *complete cases*, since this is what most statistical software packages do automatically. In a multivariate analysis, they would, for example, omit any case with a missing value on any of the variables. When the incomplete cases comprise only a small fraction of all cases (say, 5 percent or less) then case deletion may be a perfectly reasonable solution to the missing data problem. But in multivariate data sets in particular, where missing values can occur on any of the variables, the incomplete cases may often be a substantial portion of the entire dataset. If so, omitting them may cause large amounts of information to be discarded, which would clearly be very inefficient.

But the main problem with complete-case analysis is that it can lead to a serious bias in both estimation and inference unless the missing data are *missing completely at random* (see Chapter 9 and Little and Rubin, 1987, for more details). In other words, complete-case analysis implicitly assumes that the discarded cases are like a random subsample. So at the very least complete-case analysis leads to a loss, and perhaps a substantial loss in power, but worse, analyses based on just complete cases might in some cases be misleading.

So what can be done? One answer is to consider some form of *imputation*, the practice of "filling in" missing data with plausible values. At one level this will solve the missing-data problem and enable the investigator to progress normally. But from a statistical viewpoint careful consideration needs to be given to the method used for imputation; otherwise it may cause more problems than it solves. For example, imputing an observed variable mean for a variable's missing values preserves the observed sample means, but distorts the covariance structure, biasing estimated variances and covariances toward zero. On the other hand imputing predicted values from regression models tends to inflate observed correlations, biasing them away from zero. And treating imputed data as if they were "real" in estimation and inference can lead to misleading standard errors and p-values, since they fail to reflect the uncertainty due to the missing data.

The most appropriate way to deal with missing values is a procedure suggested by Rubin (1987), known as *multiple imputation*. This is a Monte Carlo technique in which the missing values are replaced by $m > 1$ simulated versions, where m is typically small (say 3–10). Each of the simulated complete datasets is analyzed by the method appropriate for the investigation at hand, and the results are later combined to produce estimates and confidence intervals that incorporate missing-data uncertainty. Details are given in Rubin (1987) and more concisely in Schafer (1999). An S-PLUS® library for multiple imputation is available; see Schimert et al. (2000). The greatest virtues of multiple imputation are its simplicity and its generality. The user may analyze the data by virtually any technique that would be appropriate if the data were complete. However, one should always bear in mind that the imputed values are not real measurements. We do not get something for nothing! And if there is a substantial proportion of the individuals with large amounts of missing data one should clearly question whether *any* form of statistical analysis is viable.

1.3 Summary Statistics for Multivariate Data

In order to summarize a multivariate data set we need to produce summaries for each of the variables separately and also to summarize the relationships between the variables. For the former we generally use *means* and *variances* (assuming that we are dealing with continuous variables), and for the latter we usually take pairs of variables at a time and look at their *covariances* or *correlations*. Population and sample versions of all of these quantities are now defined.

1.3.1 *Means*

For q variables, the population mean vector is usually represented as $\mu' = [\mu_1, \mu_2, \ldots, \mu_q]$, where

$$\mu_i = E(x_i)$$

is the population mean (or *expected value* as denoted by the E operator in the above) of the ith variable. An *estimate of* μ', based on n, q-dimensional observations, is $\bar{x}' = [\bar{x}_1, \bar{x}_2, \ldots, \bar{x}_q]$, where $\bar{x}_i$ is the sample mean of the variable x_i.

To illustrate the calculation of a mean vector we shall use the data shown in Table 1.2, which shows the heights (millimeters) and ages (years) of both partners in a sample of 10 married couples. We assume that the data are available as the data.frame huswif with variables labelled as shown in Table 1.2. The mean vector for these data can be found directly in R with the mean function and in S-PLUS by using the apply function combined with the mean function;

R: mean(huswif)
S-PLUS: apply(huswif,2,mean)

Table 1.2 Heights and Ages of Husband and Wife in 10 Married Couples

Husband age (Hage)	Husband height (Hheight)	Wife age (Wage)	Wife height (Wheight)	Husband age at first marriage (Hagefm)
49	1809	43	1590	25
25	1841	28	1560	19
40	1659	30	1620	38
52	1779	57	1540	26
58	1616	52	1420	30
32	1695	27	1660	23
43	1730	52	1610	33
47	1740	43	1580	26
31	1685	23	1610	26
26	1735	25	1590	23

The values that result are:

Hage	Hheight	Wage	Wheight	Hagefm
40.3	1728.9	38.0	1578.0	26.9

1.3.2 Variances

The vector of population variances can be represented by $\boldsymbol{\sigma}' = [\sigma_1^2, \sigma_2^2, \ldots, \sigma_q^2]$, where

$$\sigma_i^2 = E(x_i - \mu_i)^2.$$

An estimate of $\boldsymbol{\sigma}'$ based on n, q-dimensional observations is $\mathbf{s}' = [s_1^2, s_2^2, \ldots, s_q^2]$, where s_i^2 is the sample variance of x_i.

We can get the variances for the variables in the husbands and wives data set by using the sd function directly in R and again using the apply function combined with the var function in S-PLUS:

R: sd(huswif)^2
S-PLUS: apply(huswif,2,var)

to give

Hage	Hheight	Wage	Wheight	Hagefm
130.23	4706.99	164.67	4173.33	29.88

1.3.3 Covariances

The population covariance of two variables, x_i and x_j, is defined by

$$\text{Cov}(x_i, x_j) = E(x_i - \mu_i)(x_j - \mu_j).$$

If $i = j$, we note that the covariance of the variable with itself is simply its variance, and therefore there is no need to define variances and covariances independently in the multivariate case. The covariance of x_i and x_j is usually denoted by σ_{ij} (so the variance of the variable x_i is often denoted by σ_{ii} rather than σ_i^2).

With q variables, $x_1, x_2, \ldots, x_q$, there are q variances and $q(q - 1)/2$ covariances. In general these quantities are arranged in a $q \times q$ symmetric matrix, $\mathbf{\Sigma}$, where

$$\mathbf{\Sigma} = \begin{pmatrix} \sigma_{11} & \sigma_{12} & \cdots & \sigma_{1q} \\ \sigma_{21} & \sigma_{22} & \cdots & \sigma_{2q} \\ \vdots & \vdots & \vdots & \vdots \\ \sigma_{q1} & \sigma_{q2} & \cdots & \sigma_{qq} \end{pmatrix}.$$

Note that $\sigma_{ij} = \sigma_{ji}$. This matrix is generally known as the *variance–covariance matrix* or simply the *covariance matrix*. The matrix $\mathbf{\Sigma}$ is estimated by the matrix $\mathbf{S}$, given by

$$\mathbf{S} = \frac{1}{n-1} \sum_{i=1}^{n} (\mathbf{x}_i - \bar{\mathbf{x}})(\mathbf{x}_i - \bar{\mathbf{x}})'$$

where $\mathbf{x}_i' = [x_{i1}, x_{i2}, \ldots, x_{iq}]$ is the vector of observations for the ith individual. The diagonal of $\mathbf{S}$ contains the variances of each variable.

The covariance matrix for the data in Table 1.2 is obtained using the var function in both R and S-PLUS,

var(huswif)

to give the following matrix of variances (on the main diagonal) and covariances (the off diagonal elements).

	Hage	Hheight	Wage	Wheight	Hagefm
Hage	130.23	−192.19	128.56	−436.00	28.03
Hheight	−192.19	4706.99	25.89	876.44	−229.34
Wage	128.56	25.89	164.67	−456.67	21.67
Wheight	−436.00	876.44	−456.67	4173.33	−8.00
Hagefm	28.03	−229.34	21.67	−8.00	29.88

1.3.4 Correlations

The covariance is often difficult to interpret because it depends on the units in which the two variables are measured; consequently, it is often standardized by dividing by the product of the standard deviations of the two variables to give a quantity called the *correlation coefficient*, ρ_{ij}, where

$$\rho_{ij} = \frac{\sigma_{ij}}{\sqrt{\sigma_{ii}\sigma_{jj}}}.$$

The correlation coefficient lies between -1 and $+1$ and gives a measure of the *linear* relationship of the variables x_i and x_j. It is positive if high values of x_i are associated with high values of x_j and negative if high values of x_i are associated with low values of x_j. With q variables there are $q(q-1)/2$ distinct correlations which may be arranged in a $q \times q$ matrix whose diagonal elements are unity.

For sample data, the correlation matrix contains the usual estimates of the ρ's, namely Pearson's correlation coefficient, and is generally denoted by **R**. The matrix may be written in terms of the sample covariance matrix **S** as follows,

$$\mathbf{R} = \mathbf{D}^{-1/2}\mathbf{S}\mathbf{D}^{-1/2}$$

where $\mathbf{D}^{-1/2} = \text{diag}(1/s_i)$.

In most situations we will be dealing with covariance and correlation matrices of full rank, q, so that both matrices will be nonsingular (i.e., invertible).

The correlation matrix for the four variables in Table 1.2 is obtained by using the function cor in both R and S-PLUS,

cor(huswif)

to give

	Hage	Hheight	Wage	Wheight	Hagefm
Hage	1.00	-0.25	0.88	-0.59	0.45
Hheight	-0.25	1.00	0.03	0.20	-0.61
Wage	0.88	0.03	1.00	-0.55	0.31
Wheight	-0.59	0.20	-0.55	1.00	-0.02
Hagefm	0.45	-0.61	0.31	-0.02	1.00

1.3.5 Distances

The concept of distance between observations is of considerable importance for some multivariate techniques. The most common measure used in *Euclidean distance*, which for two rows, say row i and row j, of the multivariate data matrix, **X**, is defined as

$$d_{ij} = \left[\sum_{k=1}^{q}(x_{ik} - x_{jk})^2 \right]^{1/2}.$$

We can use the dist function in both R and S-PLUS to calculate these distances for the data in Table 1.2,

dis<-dist(huswif)

This can be converted into the required distance matrix by using the function dist2full given in help(dist):

```
dist2full<-function(dis) {
    n<-attr(dis,"Size")
    full<-matrix(0,n,n)
    full[lower.tri(full)]<-dis
```

```
    full+t(full)
}
dis.matrix<-dist2full(dis)
round(dis.matrix,digits=2)
```

The resulting distance matrix is

```
numeric matrix: 10 rows, 10 columns.
          [,1]    [,2]    [,3]    [,4]    [,5]    [,6]    [,7]    [,8]    [,9]   [,10]
 [1,]     0.00   52.55  154.33   60.05  257.56  135.81   82.60   69.76  128.46   79.58
 [2,]    52.55    0.00  193.17   76.57  268.35  177.15  126.16  106.58  164.15  110.28
 [3,]   154.33  193.17    0.00  147.71  206.69   56.52   75.23   92.32   32.40   84.39
 [4,]    60.05   76.57  147.71    0.00  202.60  150.88   86.35   57.81  123.83   78.39
 [5,]   257.56  268.35  206.69  202.60    0.00  255.33  222.10  202.96  206.03  211.81
 [6,]   135.81  177.15   56.52  150.88  255.33    0.00   67.61   94.42   51.24   80.87
 [7,]    82.60  126.16   75.23   86.35  222.10   67.61    0.00   33.85   55.31   39.28
 [8,]    69.76  106.58   92.32   57.81  202.96   94.42   33.85    0.00   67.68   29.98
 [9,]   128.46  164.15   32.40  123.83  206.03   51.24   55.31   67.68    0.00   54.20
[10,]    79.58  110.28   84.39   78.39  211.81   80.87   39.28   29.98   54.20    0.00
```

But this calculation of the distances ignores the fact that the variables in the data set are on different scales, and changing the scales will change the elements of the distance matrix without preserving the rank order of pairwise distances. It makes more sense to calculate the distances *after* some form of standardization. Here we shall divide each variable by its standard deviation. The necessary R code is

```
#find standard deviations of variables

std<-sd(huswif)
#use sweep function to divide columns of data matrix
#by the appropriate standard deviation
huswif.std<-sweep(huswif,2,std,FUN=''/'')
dis<-dist(huswif.std)
dis.matrix<-dist2full(dis)
round(dis.matrix,digits=2)
```

(In S-PLUS std will have to be calculated using apply and var.)
The result is the matrix given below

```
numeric matrix: 10 rows, 10 columns.
         [,1]  [,2]  [,3]  [,4]  [,5]  [,6]  [,7]  [,8]  [,9] [,10]
 [1,]    0.00  2.73  3.51  1.44  4.10  2.80  2.08  1.05  2.88  2.71
 [2,]    2.73  0.00  4.66  3.64  5.60  2.80  3.97  3.00  2.80  1.79
 [3,]    3.51  4.66  0.00  3.87  4.19  2.96  2.22  2.83  2.43  3.26
 [4,]    1.44  3.64  3.87  0.00  3.17  3.71  2.02  1.45  3.67  3.57
 [5,]    4.10  5.60  4.19  3.17  0.00  5.07  3.67  3.37  4.57  4.89
 [6,]    2.80  2.80  2.96  3.71  5.07  0.00  2.99  2.36  1.01  1.35
 [7,]    2.08  3.97  2.22  2.02  3.67  2.99  0.00  1.58  2.88  3.18
 [8,]    1.05  3.00  2.83  1.45  3.37  2.36  1.58  0.00  2.29  2.38
 [9,]    2.88  2.80  2.43  3.67  4.57  1.01  2.88  2.29  0.00  1.07
[10,]    2.71  1.79  3.26  3.57  4.89  1.35  3.18  2.38  1.07  0.00
```

In essence, in the previous section `var` and `cor` have computed *similarities* between variables, and taking `1-cor(huswif)`, for example, would give a measure of distance between the variables. More will be said about similarities and distances in Chapter 5.

1.4 The Multivariate Normal Distribution

Just as the normal distribution dominates univariate techniques, the *multivariate normal distribution* often plays an important role in some multivariate procedures. The distribution is defined explicitly in, for example, Mardia et al. (1979) and is assumed by techniques such as *multivariate analysis of variance* (MANOVA); see Chapter 7. In practice some departure from this assumption is not generally regarded as particularly serious, but it may, on occasions, be worthwhile undertaking some test of the assumption. One relatively simple possibility is to use a *probability plotting* technique. Such plots are commonly applied in univariate analysis and involve ordering the observations and then plotting them against the appropriate values of an assumed cumulative distribution function. Details are given in Display 1.1

Display 1.1
Probability Plotting

- There are two basic types of plot for comparing two probability distributions, the *probability–probability plot* and the *quantile–quantile plot*. The diagram below may be used for describing each type.

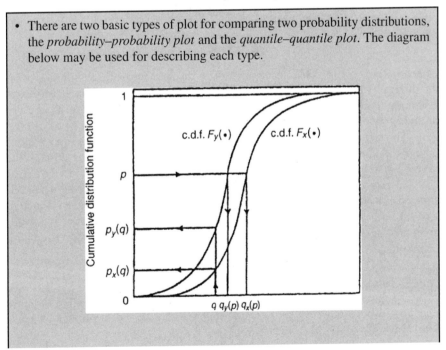

- A plot of points whose coordinates are the cumulative probabilities $(p_x(q), p_y(q))$ for different values of q is a probability–probability plot, while a plot of the points whose coordinates are the quantiles $(q_x(p), q_y(p))$ for different values of p is a quantile–quantile plot.
- An example, a quantile–quantile plot for investigating the assumption that a set of data is from a normal distribution would involve plotting the ordered sample values $y_{(1)}, y_{(2)}, K, y_{(n)}$ against the quantiles of a standard normal distribution, $\Phi^{-1}[p(i)]$, where usually

$$p_i = \frac{i - \frac{1}{2}}{n} \quad \text{and} \quad \Phi(x) = \int_{-\infty}^{x} \frac{1}{\sqrt{2\pi}} e^{-\frac{1}{2}u^2} \, du.$$

- This is usually known as a *normal probability plot*.

For multivariate data such plots may be used to examine each variable separately, although marginal normality does not necessarily imply that the variables follow a multivariate normal distribution. Alternatively (or additionally), the multivariate observation might be converted to a single number in some way before plotting. For example, in the specific case of assessing a data set for multivariate normality, each q-dimensional observation $\mathbf{x}_i$, could be converted into a *generalized distance* (essentially *Mahalanobis distance*—see Everitt and Dunn, 2001), d_i^2 giving a measure of the distance of the particular observation from the mean vector of the complete sample, $\bar{\mathbf{x}}$; d_i^2 is calculated as

$$d_i^2 = (\mathbf{x}_i - \bar{\mathbf{x}})' \mathbf{S}^{-1} (\mathbf{x}_i - \bar{\mathbf{x}}),$$

where $\mathbf{S}$ is the sample covariance matrix. This distance measure takes into account the different variances of the variables and the covariances of pairs of variables. If the observations do arise from a multivariate normal distribution, then these distances have, approximately, a *chi-squared distribution* with q degrees of freedom. So, plotting the ordered distances against the corresponding quantiles of the appropriate chi-square distribution should lead to a straight line through the origin.

First, let us consider some probability plots of a set of multivariate data constructed to have a multivariate normal distribution. We shall first use the R function mvrnorm (the MASS library will need to be loaded to make the function available) and the S-PLUS function rmvnorm to create 200 bivariate observation with correlation coefficient 0.5;

R:
```
#load MASS library
library(MASS)
#set seed for random number generation to get the same plots
set.seed(1203)
X<-mvrnorm(200,mu=c(0,0),Sigma=matrix(c(1,0.5,0.5,1.0),ncol=2))
```

S-PLUS:

```
set.seed(1203)
X<-rmvnorm(200,rho=0.5,d=2)
```

(The data generated by R and S-PLUS will not be the same. The results below are those obtained from the data generated by R.)

The probability plots for the individual variables are obtained using the following R and S-PLUS code:

```
#set up plotting area to take two side-by-side plots
par(mfrow=c(1,2))
qqnorm(X[,1],ylab="Ordered observations")
qqline(X[,1])
qqnorm(X[,2],ylab="Ordered observations")
qqline(X[,2])
```

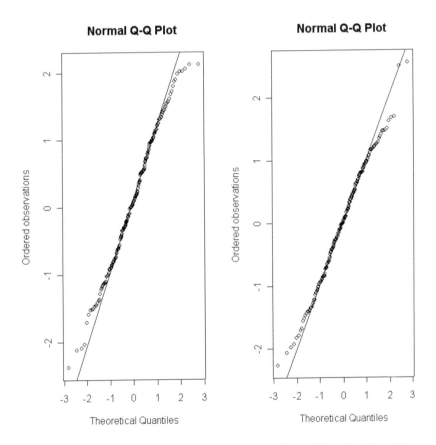

Figure 1.1 Probability plots for both variables in a generated set of bivariate data with $n = 200$ and a correlation of 0.5.

```
#qqnorm produces the required plot and qqline the line
#corresponding to a normal distribution
```

The resulting plots are shown in Figure 1.1. Neither probability plot gives any indication of a departure from linearity as we would expect.

The chi-square plot for both variables simultaneously can be found using the function chisplot given on the website mentioned in the preface. The required code is

```
par(mfrow=c(1,1))
chisplot(X)
```

Here the result appears in Figure 1.2. The plot is approximately linear, although some points do depart a little from the line.

If we now transform the previously generated data by simply taking the log of the absolute values of the generated data and then redo the previous plots, the results are shown in Figures 1.3 and 1.4. In each plot, there is a very clear departure from linearity, indicating the non-normality of the data.

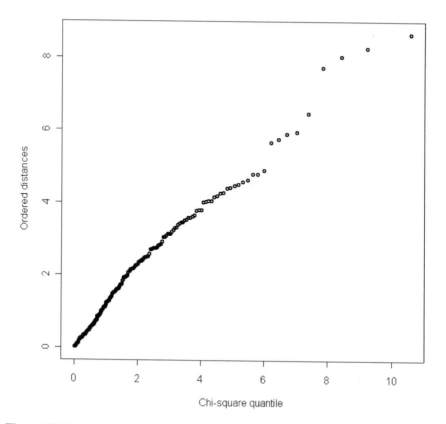

Figure 1.2 Chi-square probability plot of generated bivariate data.

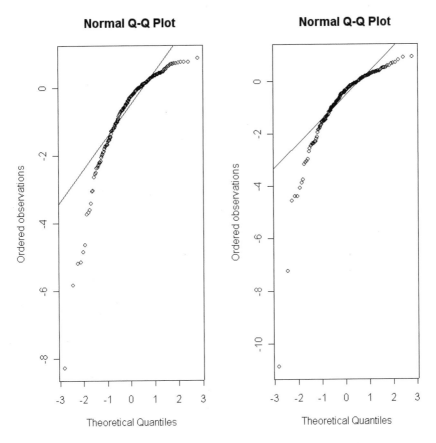

Figure 1.3 Probability plots of each variable in the transformed bivariate data.

1.5 The Aims of Multivariate Analysis

It is helpful to recognize that the analysis of data involves two separate stages. The first, particularly in new areas of research, involves *data exploration* in an attempt to recognize any nonrandom pattern or structure requiring explanation. At this stage, finding the question is often of more interest than seeking the subsequent answer. The aim of this part of the analysis being to generate possible interesting hypotheses for further study. (This activity is now often described as *data mining*.) Here, formal models designed to yield specific answers to rigidly defined questions are not required. Instead, methods are sought that allow possibly unanticipated patterns in the data to be detected, opening up a wide range of competing explanations. Such techniques are generally characterized by their emphasis on the importance of visual displays and graphical representations and by the lack of any associated stochastic model, so that questions of the statistical significance of the results are hardly ever of much importance.

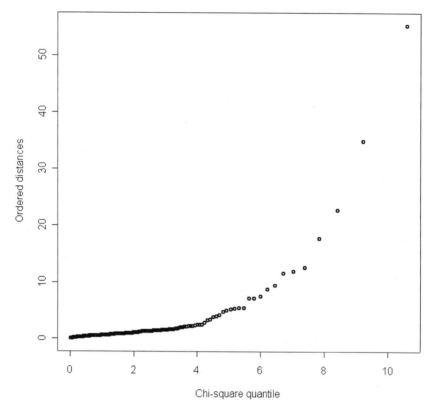

Figure 1.4 Chi-square plot of generated bivariate data after transformation.

A confirmatory analysis becomes possible after a research worker has some well-defined hypothesis in mind. It is here that some type of statistical significance test might be considered. Such tests are well known and, although their misuse has often brought them into some disrepute, they remain of considerable importance.

In this text, Chapters 2–6 describe techniques that are primarily exploratory, and Chapters 7–9 techniques that are largely confirmatory, but this division should not be regarded as much more than a convenient arrangement of the material to be presented, since any sensible investigator will realize the need for exploratory and confirmatory techniques, and many methods will often be useful in both roles. Perhaps attempts to rigidly divide data analysis into exploratory and confirmatory parts have been misplaced, and what is really important is that research workers should have a flexible and pragmatic approach to the analysis of their data, with sufficient expertise to enable them to choose the appropriate analytical tool and use it correctly. The choice of tool, of course, depends on the aims or purpose of the analysis.

Most of this text is written from the point of view that there are no rules or laws of scientific inference—that is, "anything goes" (Feyerabend, 1975). This implies that we see both exploratory and confirmatory methods as two sides of the same coin. We see both methods as essentially tools for data exploration rather than as formal decision-making procedures. For this reason we do not stress the values of significance levels, but merely use them as criteria to guide a modelling process (using the term "modelling" as a method or methods of describing the structure of a data set). We believe that in scientific research it is the skillful interpretation of evidence and subsequent development of hunches that are important, rather than a rigid adherence to a formal set of decision rules associated with significance tests (or any other criteria, for that matter). One aspect of the scientific method, however, which we do not discuss in any detail, but which is the vital component in testing the theories that come out of our data analyses, is replication. It is clearly unsafe to search for a pattern in a given data set and to "confirm" the existence of such a pattern using the same data set. We need to validate our conclusions using further data. At this point our subsequent analysis might become truly confirmatory.

1.6 Summary

Most data collected in the social sciences and other disciplines are multivariate. To fully understand most such data sets the variables need to be analyzed simultaneously. The remainder of this book is concerned with methods that have been developed to make this possible, and to help discover any patterns or structure in the data that may have important implications in uncovering the data's message.

2
Looking at Multivariate Data

2.1 Introduction

Most of the chapters in this book are concerned with methods for the analysis of multivariate data, which are based on relatively complex mathematics. This chapter, however, is not. Here we look at some relatively simple graphical procedures and there is no better software for producing graphs than R and S-PLUS®.

According to Chambers et al. (1983) "there is no statistical tool that is as powerful as a well-chosen graph." Certainly graphical presentation has a number of advantages over tabular displays of numerical results, not the least of which is creating interest and attracting the attention of the viewer. Graphs are very popular. It has been estimated that between 900 billion (9×10^{11}) and 2 trillion (2×10^{12}) images of statistical graphics are printed each year. Perhaps one of the main reasons for such popularity is that graphical presentation of data often provides the vehicle for discovering the unexpected; the human visual system is very powerful in detecting patterns, although the following caveat from the late Carl Sagan should be kept in mind.

> Humans are good at discerning subtle patterns that are really there, but equally so at imagining them when they are altogether absent.

During the last two decades a wide variety of new methods for displaying data graphically have been developed. These will hunt for special effects in data, indicate outliers, identify patterns, diagnose models and generally search for novel and perhaps unexpected phenomena. Large numbers of graphs may be required, and computers are generally needed to generate them for the same reasons they are used for numerical analyses, namely, they are fast and they are accurate.

So, because the machine is doing the work, the question is no longer "Shall we plot?" but rather "What shall we plot?" There are many exciting possibilities including, dynamic graphics (see Cleveland and McGill, 1987), but graphical exploration of data usually begins with some simpler, well-known methods. Univariate marginal views of multivariate data might, for example, be obtained using *histograms*, *stem-and-leaf plots*, or *box plots*. More important for exploring multivariate data are plots that allow the relationships between variables to be assessed. Consequently we begin our discussion of graphics with the ubiquitous scatterplot.

2.2 Scatterplots and Beyond

The simple xy scatterplot has been in used since at least the eighteenth century and has many virtues. Indeed, according to Tufte (1983):

> The relational graphic—in its barest form the scatterplot and its variants—is the greatest of all graphical designs. It links at least two variables, encouraging and even imploring the viewer to assess the possible causal relationship between the plotted variables. It confronts causal theories that x causes y with empirical evidence as to the actual relationship between x and y.

To illustrate the use of the scatterplot and the other techniques to be discussed in subsequent sections we shall use the data shown in Table 2.1. These data give

Table 2.1 Air Pollution Data for Regions in the United States

Region	Rainfall	Educ	Popden	Nonwhite	NOX	SO2	Mortality
AkronOH	36	11.4	3243	8.8	15	59	921.9
AlbanyNY	35	11.0	4281	3.5	10	39	997.9
AllenPA	44	9.8	4260	0.8	6	33	962.4
AtlantGA	47	11.1	3125	27.1	8	24	982.3
BaltimMD	43	9.6	6441	24.4	38	206	1071.0
BirmhmAL	53	10.2	3325	38.5	32	72	1030.0
BostonMA	43	12.1	4679	3.5	32	62	934.7
BridgeCT	45	10.6	2140	5.3	4	4	899.5
BufaloNY	36	10.5	6582	8.1	12	37	1002.0
CantonOH	36	10.7	4213	6.7	7	20	912.3
ChatagTN	52	9.6	2302	22.2	8	27	1018.0
ChicagIL	33	10.9	6122	16.3	63	278	1025.0
CinnciOH	40	10.2	4101	13.0	26	146	970.5
ClevelOH	35	11.1	3042	14.7	21	64	986.0
ColombOH	37	11.9	4259	13.1	9	15	958.8
DallasTX	35	11.8	1441	14.8	1	1	860.1
DaytonOH	36	11.4	4029	12.4	4	16	936.2
DenverCO	15	12.2	4824	4.7	8	28	871.8
DetrotMI	31	10.8	4834	15.8	35	124	959.2
FlintMI	30	10.8	3694	13.1	4	11	941.2
FtwortTX	31	11.4	1844	11.5	1	1	891.7
GrndraMI	31	10.9	3226	5.1	3	10	871.3
GrnborNC	42	10.4	2269	22.7	3	5	971.1
HartfdCT	43	11.5	2909	7.2	3	10	887.5

(Continued)

Table 2.1 (*Continued*)

Region	Rainfall	Educ	Popden	Nonwhite	NOX	SO2	Mortality
HoustnTX	46	11.4	2647	21.0	5	1	952.5
IndianIN	39	11.4	4412	15.6	7	33	968.7
KansasMO	35	12.0	3262	12.6	4	4	919.7
LancasPA	43	9.5	3214	2.9	7	32	844.1
LosangCA	11	12.1	4700	7.8	319	130	861.8
LouisvKY	30	9.9	4474	13.1	37	193	989.3
MemphsTN	50	10.4	3497	36.7	18	34	1006.0
MiamiFL	60	11.5	4657	13.5	1	1	861.4
MilwauWI	30	11.1	2934	5.8	23	125	929.2
MinnplMN	25	12.1	2095	2.0	11	26	857.6
NashvlTN	45	10.1	2082	21.0	14	78	961.0
NewhvnCT	46	11.3	3327	8.8	3	8	923.2
NeworlLA	54	9.7	3172	31.4	17	1	1113.0
NewyrkNY	42	10.7	7462	11.3	26	108	994.6
PhiladPA	42	10.5	6092	17.5	32	161	1015.0
PittsbPA	36	10.6	3437	8.1	59	263	991.3
PortldOR	37	12.0	3387	3.6	21	44	894.0
ProvdcRI	42	10.1	3508	2.2	4	18	938.5
ReadngPA	41	9.6	4843	2.7	11	89	946.2
RichmdVA	44	11.0	3768	28.6	9	48	1026.0
RochtrNY	32	11.1	4355	5.0	4	18	874.3
StLousMO	34	9.7	5160	17.2	15	68	953.6
SandigCA	10	12.1	3033	5.9	66	20	839.7
SanFranCA	18	12.2	4253	13.7	171	86	911.7
SanJosCA	13	12.2	2702	3.0	32	3	790.7
SeatleWA	35	12.2	3626	5.7	7	20	899.3
SpringMA	45	11.1	1883	3.4	4	20	904.2
SyracuNY	38	11.4	4923	3.8	5	25	950.7
ToledoOH	31	10.7	3249	9.5	7	25	972.5
UticaNY	40	10.3	1671	2.5	2	11	912.2
WashDC	41	12.3	5308	25.9	28	102	968.8
WichtaKS	28	12.1	3665	7.5	2	1	823.8
WilmtnDE	45	11.3	3152	12.1	11	42	1004.0
WorctrMA	45	11.1	3678	1.0	3	8	895.7
YorkPA	42	9.0	9699	4.8	8	49	911.8
YoungsOH	38	10.7	3451	11.7	13	39	954.4

Data assumed available as dataframe `airpoll` with variable names as indicated.

information on 60 U.S. metropolitan areas (McDonald and Schwing, 1973; Henderson and Velleman, 1981). For each area the following variables have been recorded:

1. *Rainfall:* mean annual precipitation in inches
2. *Education:* median school years completed for those over 25 in 1960
3. *Popden:* population/mile^2 in urbanized area in 1960
4. *Nonwhite:* percentage of urban area population that is nonwhite
5. *NOX:* relative pollution potential of oxides of nitrogen
6. *SO2:* relative pollution potential of sulphur dioxide
7. *Mortality:* total age-adjusted mortality rate, expressed as deaths per 100,000

One of the questions about these data might be "How is sulphur dioxide pollution related to mortality?" A first step in answering the question would be to examine a scatterplot of the two variables. Here, in fact, we will produce four versions of the basic scatterplot using the following R and S-PLUS code (we assume that the data are available as the data frame `airpoll` with variable names as above):

```
attach(airpoll)
#set up plotting area to take four graphs
par(mfrow=c(2,2,))
par(pty="s")
plot(SO2,Mortality,pch=1,lwd=2)
title("(a)",lwd=2)
plot(SO2,Mortality,pch=1,lwd=2)
#add regression line
abline(lm(Mortality~SO2),lwd=2)
title("(b)",lwd=2)
#jitter data
airpoll1<-jitter(cbind(SO2,Mortality))
plot(airpoll1[,1],airpoll1[,2],xlab="SO2",ylab="Mortality",
    pch=1,lwd=2)
title("(c)",lwd=2)
plot(SO2,Mortality,pch=1,lwd=2)
#add rug plots
rug(jitter(SO2),side=1)
rug(jitter(Mortality),side=2)
title("(d)",lwd=2)
```

Figure 2.1(a) shows the scatterplot of *Mortality* against *SO2*. Figure 2.1(b) shows the same scatterplot with the addition of the simple linear regression fit of *Mortality* on *SO2*. Both plots suggest a possible link between increasing sulphur dioxide level and increasing mortality.

Although not a real problem here, scatterplots in which there are many points often suffer from overplotting. The problem can be overcome, partially at least, by "jittering" the data, that is, adding a small amount of noise to each observation

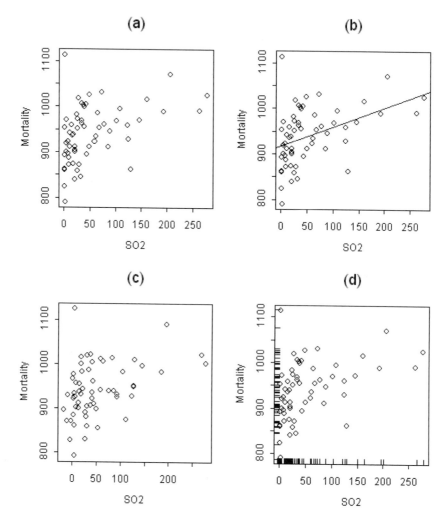

Figure 2.1 (**a**) Scatterplot for Mortality against SO2; (**b**) scatterplot of Mortality against SO2 with added linear regression fit; (**c**) jittered scatterplot of Mortality against SO2; (**d**) scatterplot of Mortality against SO2 with information about marginal distributions of the two variables added.

before plotting (see Chambers et al., 1983, for details). Figure 2.1(c) shows the scatterplot in Figure 2.1(a) after jittering. Finally, in Figure 2.1(d), the bivariate scatter of the two variables is framed with a display of the marginal distribution of each variable. Plotting marginal and joint distributions together is usually good data analysis practice.

With these data it might be useful to label the scatterplot with the names of the regions involved. These names are rather long, and if used as they are would

lead to a rather "messy" plot; consequently we shall use the R and S-PLUS function
abbreviate to shorten them before plotting using the code:

```
names<-abbreviate(row.names(airpoll))
plot(SO2,Mortality,lwd=2,type="n")
text(SO2,Mortality,labels=names,lwd=2)
```

Figure 2.2 highlights some regions with odd combinations of pollution and
mortality values. For example, *nwLA* has almost zero SO_2 value, but very high
mortality. Perhaps this is a garden suburb where people go to retire?

In Figure 2.1(b) a simple linear regression fit was added to the *Mortality/SO2*
scatterplot. This addition is often very useful for assessing the relationship between
the two variables more accurately. Even more useful is to add both the linear regression fit and a *locally weighted regression* or *lowest* fit to the scatterplot. Such fits
are described in detail in Cleveland (1979), but essentially they are designed to
use the data themselves to suggest the type of fit needed. The model assumed
is that

$$y_i = g(x_i) + \epsilon_i,$$

where g is a "smooth" function and the ϵ_i are random variables with zero mean and
constant variance. Fitted values, $\hat{y}_i$, are used to estimate $g(x_i)$ at each x_i by fitting
polynomials using weighted least squares, with large weights for points close to

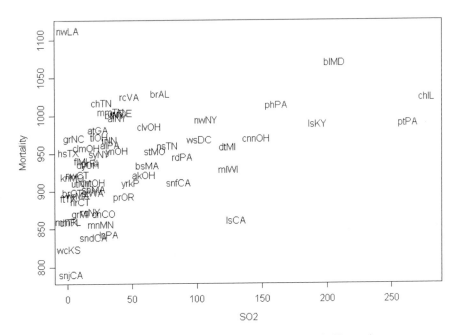

Figure 2.2 Scatterplot of mortality against SO_2 with points labeled by region name.

x_i and small weights otherwise. The degree of "smoothness" of the fitted curve can be controlled by a particular parameter during the fitting process. Examining a scatterplot that includes a locally weighted regression fit can often be a useful antidote to the thoughtless fitting of straight lines with least squares.

To illustrate the use of lowest fits we return to the air pollution data and again concentrate on the two variables, *SO2* and *Mortality*. The following R and S-PLUS code produces a scatterplot with some information about marginal distributions that also includes both a linear regression and a locally weighted regression fit:

```
#set up plotting area for scatterplot
par(fig=c(0,0.7,0,0.7))
plot(SO2,Mortality,lwd=2)
#add regression line
abline(lm(Mortality~SO2),lwd=2)
#add locally weighted regression fit
lines(lowess(SO2,Mortality),lwd=2)
#set up plotting area for histogram
par(fig=c(0,0.7,0.65,1),new=TRUE)
hist(SO2,lwd=2)
#set up plotting area for boxplot
par(fig=c(0.65,1,0,0.7),new=TRUE)
boxplot(Mortality,lwd=2)
```

The resulting diagram is shown in Figure 2.3. Here, apart from a small "wobble" for sulphur dioxide values 0 to 100, the linear fit and the locally weighted fit are very similar.

2.2.1 *The Convex Hull of Bivariate Data*

Scatterplots are often used in association with the calculation of the correlation coefficient of two variables. Outliers, for example, can often considerably distort the value of a correlation coefficient, and a scatterplot may help to identify the offending observations, which might then be excluded from the calculation. Another approach that allows *robust estimation* of the correlation is *convex hull trimming*. The convex hull of a set of bivariate observations consists of the vertices of the smallest convex polyhedron in variable space within which, or on which, all data points lie. Removal of the points lying on the convex hull can eliminate isolated outliers without disturbing the general shape of the bivariate distribution. A robust estimate of the correlation coefficient results from using the remaining observations.

Let's see how the convex hull approach works with our *Mortality/SO2* scatterplot. We can calculate the correlation coefficient of the two variables using all the observations from the R and S-PLUS instruction:

```
cor(SO2, Mortality)
```

giving a value of 0.426.

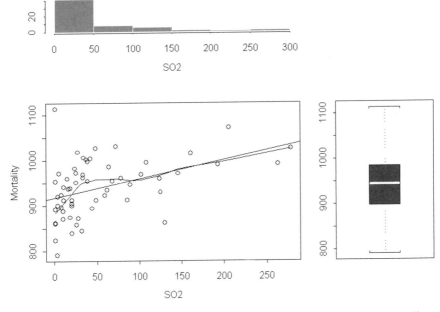

Figure 2.3 Scatterplot of mortality against SO$_2$ with added linear regression and locally weighted regression fits and marginal distribution information.

Now we can find the convex hull of the data and, for interest, show it on a scatterplot of the two variables using the following R and S-PLUS code:

```
#find points defining convex hull
hull<-chull(SO2,Mortality)
plot(SO2,Mortality,pch=1)
#plot and shade convex hull
polygon(SO2[hull],Mortality[hull],density=15,angle=30)
```

The result is shown in Figure 2.4.

To calculate the correlation coefficient after removal of the points defining the convex hull requires the instruction

```
cor(SO2[-hull],Mortality[-hull])
```

The resulting value of the correlation is now 0.438. In this case the change in the correlation after removal of the points defining the convex hull is very small, surprisingly small, given that some of the defining observations are relatively remote from the body of the data.

2.2.2 *The Chiplot*

Although the scatterplot is a primary data-analytic tool for assessing the relationship between a pair of continuous variables, it is often difficult to judge whether or not

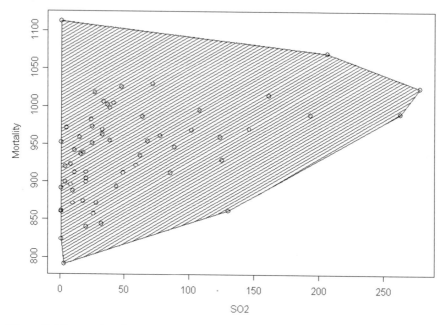

Figure 2.4 Scatterplot of mortality against SO_2 showing convex hull of the data.

the variables are independent. A random scatter of points may be hard for the human eye to judge. Consequently, it is often helpful to augment the scatterplot with an auxiliary display in which independence is itself manifested in a characteristic manner. The *chi-plot* suggested by Fisher and Switzer (1985, 2001) is designed to address the problem. The essentials of this type of plot are described in Display 2.1.

Display 2.1
The Chi-Plot

- A chi-plot is a scatterplot of the pairs

$$(\lambda_i, \chi_i), |\lambda_2| < 4\left\{\frac{1}{n-1} - \frac{1}{2}\right\}^2,$$

where

$$\chi_i = (H_i - F_i G_i)/\{F_i(1 - F_i)G_i(1 - G_i)\}^{1/2}$$

$$\lambda_i = 4S_i \max\left\{\left(F_i - \frac{1}{2}\right)^2, \left(G_i - \frac{1}{2}\right)^2\right\}$$

and

$$H_i = \sum_{j \neq i} I(x_j \leq x_i, y_j \leq y_i)/(n-1)$$

$$F_i = \sum_{j \neq i} I(x_j \leq x_i)/(n-1)$$

$$G_i = \sum_{j \neq i} I(y_j \leq y_i)/(n-1)$$

$$S_i = \text{sign}\left\{\left(F_i - \frac{1}{2}\right)\left(G_i - \frac{1}{2}\right)\right\}$$

where sign (x) is $+1$ if x is positive, 0 if x is zero, and -1 if x is negative; $I(A)$ is the indicator function for the event A, that is, if A is true $I(A) = 1$, if A is not true, $I(A) = 0$.
- When the two variables are independent, the points in a chi-plot will be scattered about a central region. When they are related, the points will tend to lie outside this central region. See the example in the text.

An R and S-PLUS function for producing chi-plots, the `chiplot` is given on the website mentioned in the Preface. To illustrate the chi-plot we shall apply it to the *Mortality* and *SO2* variables of the air pollution data using the code

```
chiplot(SO2,Mortality,vlabs=c("SO2","Mortality"))
```

The result is Figure 2.5 which shows the scatterplot of *Mortality* plotted against *SO2* alongside the corresponding chi-plot. Departure from independence is indicated in the latter by a lack of points in the horizontal band indicated on the plot. Here there is a clear departure since there are very few of the observations in this region.

2.2.3 *The Bivariate Boxplot*

A further helpful enhancement to the scatterplot is often provided by the two-dimensional analogue of the boxplot for univariate data, known as the *bivarate boxplot* (Goldberg and Iglewicz, 1992). This type of boxplot may be useful in indicating the distributional properties of the data and in identifying possible outliers. The bivariate boxplot is based on calculating "robust" measures of location, scale, and correlation. It consists essentially of a pair of concentric ellipses, one of which (the "hinge") includes 50% of the data and the other (called the "fence") which delineates potential troublesome outliers. In addition, resistant regression lines of both y on x and x on y are shown, with their intersection showing the bivariate location estimator. The acute angle between the regression lines will be small

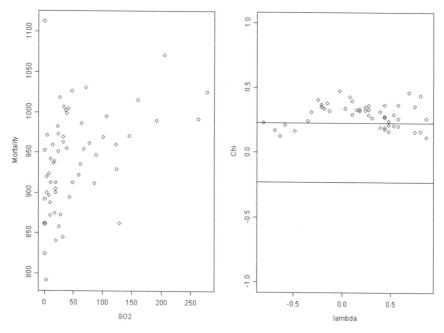

Figure 2.5 Chi-plot of *Mortality* and *SO2*.

for a large absolute value of correlations and large for a small one. Details of the construction of a bivarate boxplot are as given in Display 2.2:

Display 2.2
Constructing a Bivariate Boxplot

- The bivariate boxplot is the two-dimensional analogue of the familiar boxplot for univariate data and consists of a pair of concentric ellipses, the "hinge" and the "fence."
- To draw the elliptical fence and hinge, location (T_x^*, T_y^*), scale (S_x^*, S_y^*), and correlation (R^*) estimators are needed, in addition to a constant D that regulates the distance of the fence from the hinge. In general $D = 7$ is recommended since this corresponds to an approximate 99% confidence bound on a single observation.
- In general, robust estimators of location, scale, and correlation are recommended since they are better at handling data with outliers or with density or shape differing moderately from the elliptical bivariate normal. Goldberg and Iglewicz (1992) discuss a number of possibilities.

- To draw the bivariate boxplot, first calculate the median E_m and the maximum E_{max} of the standardized errors, E_i, which are essentially the generalized distances of each point from the centre (T_x^*, T_y^*). Specifically, the E_i are defined by

$$E_i = \sqrt{\frac{X_{si}^2 + Y_{si}^2 - 2R^* X_{si} Y_{si}}{1 - R^{*2}}},$$

where $X_{si} = (X_i - T_x^*)/S_x^*$ is the standardized X_i value and Y_{si} is similarly defined.
- Then

$$E_m = \text{median } \{E_i : i = 1, 2, K, n\}$$

and

$$E_{max} = \text{maximum } \{E_i : E_i^2 < DE_m^2\}.$$

- To draw the hinge, let

$$R_1 = E_m \sqrt{\frac{1 + R^*}{2}}, \quad R_2 = E_m \sqrt{\frac{1 - R^*}{2}}.$$

- For $\theta = 0$ to 360 in steps of 2, 3, 4, or 5 degrees, let

$$\Theta_1 = R_1 \cos \theta,$$
$$\Theta_2 = R_2 \sin \theta,$$
$$X = T_x^* + (\Theta_1 + \Theta_2) S_x^*,$$
$$Y = T_y^* + (\Theta_1 - \Theta_2) S_y^*.$$

- Finally, plot X, Y.

To illustrate the use of a bivariate boxplot we shall again use the *SO2* and *Mortality* scatterplot. An R and S-PLUS function, `bivbox`, for constructing and plotting the boxplot is given on the website (see Preface) and can be used as follows,

```
bivbox(cbind(SO2,Mortality),xlab="SO2",ylab="Mortality"))
```

to give the diagram shown in Figure 2.6.

In Figure 2.6 robust estimators of scale and location have been used and the diagram suggests that there are five outliers in the data. To use the nonrobust

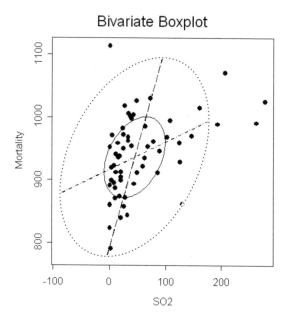

Figure 2.6 Bivariate boxplot of *SO2* and *Mortality* (robust estimators of location, scale, and correlation).

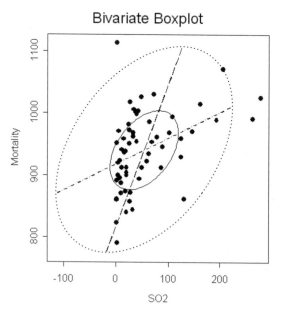

Figure 2.7 Bivariate boxplot of *SO2* and *Mortality* (nonrobust estimators).

estimators, that is, the usual means, variances, and correlation coefficient, the necessary code is

```
bivbox(cbind(SO2,Mortality),xlab="SO2",ylab="Mortality",
   method="O")
```

The resulting diagram is shown in Figure 2.7. Now only three outliers are identified. In general the use of the robust estimator version of the `bivbox` function is recommended.

2.3 Estimating Bivariate Densities

Often the aim in examining scatterplots is to identify regions where there are high or low densities of observations, "clusters," or to spot outliers. But humans are not particularly good at visually examining point density, and it is often a very helpful aid to add some type of *bivariate density estimate* to the scatterplot. In general a nonparametric estimate is most useful since we are unlikely, in most cases, to want to assume some particular parametric form such as the bivariate normality. There is now a considerable literature on density estimation; see, for example, Silverman (1986) and Wand and Jones (1995). Basically, density estimates are "smoothed" two-dimensional histograms. A brief summary of the mathematics of bivariate density estimation is given in Display 2.3.

Display 2.3
Estimating Bivariate Densities

- The data set whose underlying density is to be estimated is $\mathbf{X}_1, \mathbf{X}_2, L, \mathbf{X}_n$.
- The bivariate kernel density estimator with kernel K and window width h is defined by

$$\hat{f}(\mathbf{x}) = \frac{1}{nh^2} \sum_{i=1}^{n} K\left\{ \frac{1}{h}(\mathbf{x} - \mathbf{X}_i) \right\}.$$

- The kernel function $K(\mathbf{x})$ is a function, defined for bivariate $\mathbf{x}$, satisfying

$$\int K(\mathbf{x})d\mathbf{x} = 1.$$

- Usually $K(\mathbf{x})$ will be a radially symmetric unimodal probability density function, for example, the standard bivariate normal density function:

$$K(\mathbf{x}) = \frac{1}{2\pi} \exp\left(-\frac{1}{2}\mathbf{x}'\mathbf{x} \right).$$

Let us look at a simple two-dimensional histogram of the *Mortality/SO2* observations found and then displayed as a perspective plot by using the S-PLUS code

```
h2d<-hist2d(SO2,Mortality)
persp(h2d,xlab="SO2",ylab="Mortality",zlab="Frequency")
```

The result is shown in Figure 2.8. The density estimate given by the histogram is really too rough to be useful. (The function `hist2d` appears to be unavailable in R, but this is of little consequence since, in practice, unsmoothed two-dimensional histograms are of little use.)

Now we can use the R and S-PLUS function `bivden` given on the website to find a smoother estimate of the bivariate density of *Mortality* and *SO2* and to then display the estimated density as both a contour and perspective plot. The necessary code is

```
#get bivariate density estimates using a normal kernel
den1<-bivden(SO2,Mortality)
#construct a perspective plot of the density values
persp(den1$seqx,den1$seqy,den1$den,xlab="SO2",
  ylab="Mortality",
zlab="Density",lwd=2)
#
plot(SO2,Mortality)
#add a contour plot of the density values to the scatterplot
contour(den1$seqx,den1$seqy,den1$den,lwd=2,nlevels=20,add=T)
```

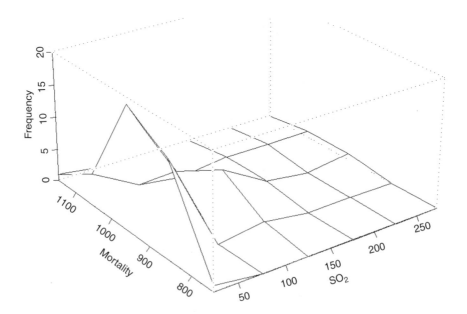

Figure 2.8 Two-dimensional histogram of *Mortality* and *SO2*.

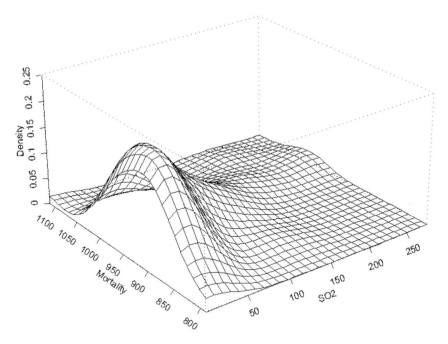

Figure 2.9 Perspective plot of estimated bivariate density of *Mortality* and *SO2*.

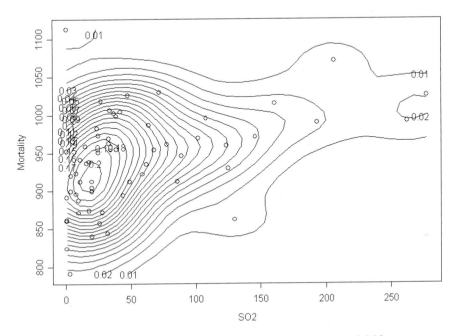

Figure 2.10 Contour plot of estimated bivariate density of *Mortality* and *SO2*.

The results are shown in Figures 2.9 and 2.10. Both plots give a clear indication of the skewness in the bivariate density of the two variables. (The diagrams shown result from using S-PLUS; those from R are a little different.)

In R the `bkde2D` function from the *KernSmooth* library might also be used to provide bivariate density estimates; see Exercise 2.7.

2.4 Representing Other Variables on a Scatterplot

The scatterplot can only display two variables. But there have been a number of suggestions as to how extra variables may be included. In this section we shall illustrate one of these, the *bubbleplot*, in which three variables are displayed. Two variables are used to form the scatterplot itself, and then the values of the third variable are represented by circles with radii proportional to these values and centered on the appropriate point in the scatterplot. To illustrate the bubbleplot we shall use the three variables, *SO2*, *Rainfall*, and *Mortality* from the air pollution data. The R and S-PLUS code needed to produce the required bubble plot is

```
plot(SO2,Mortality,pch=1,lwd=2,ylim=c(700,1200),
    xlim=c(-5,300))
#add circles to scatterplot
symbols(SO2,Mortality,circles=Rainfall,inches=0.4,add=TRUE,
    lwd=2)
```

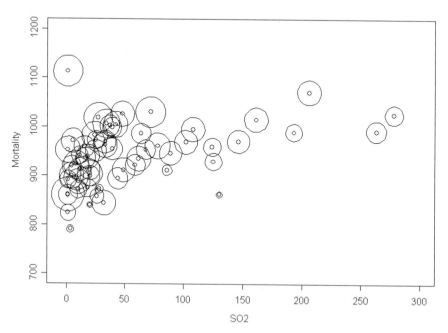

Figure 2.11 Bubbleplot of *Mortality* and *SO2* with *Rainfall* represented by radii of circles.

The resulting diagram is shown in Figure 2.11. Two particular observations to note are the one with high mortality and rainfall but very low sulphur dioxide level (NworlLA) and the one with relatively low mortality and low rainfall but moderate sulphur dioxide level (losangCA).

2.5 The Scatterplot Matrix

There are seven variables in the air pollution data which between them generate 21 possible scatterplots, and it is very important that the separate plots are presented in the best way to an in overall comprehension and understanding of the data. The *scatterplot matrix* is intended to accomplish this objective.

A scatterplot matrix is defined as a square, symmetric grid of bivariate scatterplots. The grid has q rows and columns, each one corresponding to a different variable. Each of the grid's cells shows a scatterplot of two variables. Variable j is plotted against variable i in the ijth cell, and the same variables appear in cell ji with the x- and y-axes of the scatterplots interchanged. The reason for including both the upper and lower triangles of the grid, despite the seeming redundancy, is that it enables a row and a column to be visually scanned to see one variable against all others, with the scales for the one variable lined up along the horizontal or the vertical.

To produce the basic scatterplot matrix of the air pollution variable we can use the `pairs` function in both R and S-PLUS

```
pairs(airpoll)
```

The result is Figure 2.12. The plot highlights that many pairs of variables in the air pollution data appear to be related in a relatively complex fashion, and that there are some potentially troublesome outliers in the data.

Rather than having variable labels on the main diagonal as in Figure 2.10, we may like to have some graphical representation of the marginal distribution of the corresponding variable, for example, a histogram. And here is a convenient point in the discussion to illustrate briefly the "click-and-point" features of the S-PLUS GUI since these can be useful in some situations, although for serious work the command line approach used up to now, and in most of the remainder of the book, is to be recommended. So to construct the required plot:

- Click on **Graph** in the toolbar;
- Select **2D plot**;
- In Axes Type highlight **Matrix**;
- Click **OK**;
- In **Scatterplot Matrix Dialogue** select `airpoll` as data set;
- Highlight all variables names in **x-column slot**;
- Check **Line/Histogram** tab;
- Check **Draw Histogram**;
- Click on **OK**.

The resulting diagram appears in Figure 2.13.

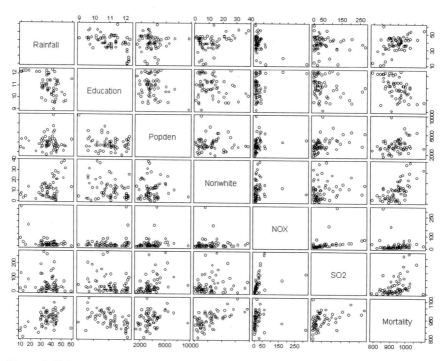

Figure 2.12 Scatterplot matrix of air pollution data.

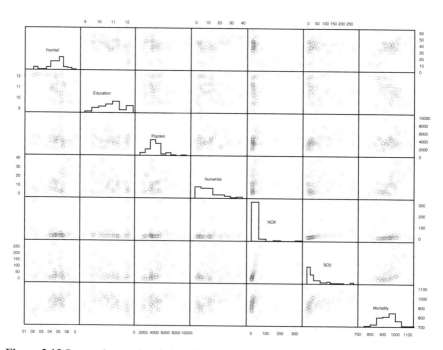

Figure 2.13 Scatterplot matrix of air pollution data showing histograms of each variable on the main diagonal.

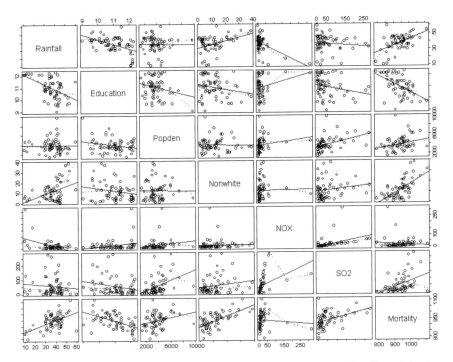

Figure 2.14 Scatterplot matrix of air pollution data showing linear and locally weighted regression fits on each panel.

Previously in this chapter we looked at a variety of ways in which individual scatterplots can be enhanced to make them more useful. These enhancements can, of course, also be used on each panel of a scatterplot matrix. For example, we can add linear and locally weighted regression fits to the air pollution diagram using the following code in either R or S-PLUS

```
pairs(airpoll,panel=function(x,y) {abline(lsfit(x,y)$coef,
                                    lwd=2)
                                    lines(lowess(x,y),lty=2,
                                    lwd=2)
                                    points(x,y)})
```

to give Figure 2.14. Other possibilities for enhancing the panels of a scatterplot matrix are considered in the exercises.

2.6 Three-Dimensional Plots

In S-PLUS there are a variety of three-dimensional plots that can often be usefully applied to multivariate data. We will illustrate some of the possibilities using once

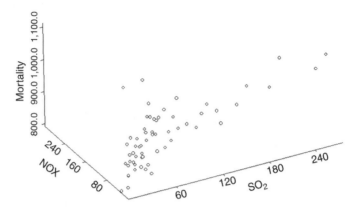

Figure 2.15 Three-dimensional plot of *SO2*, *NOX*, and *Mortality*.

again the air pollution data. To begin we will construct a simple three-dimensional plot of *SO2*, *NOX*, and *Mortality* again using the S-PLUS GUI:

- Click **Graph** on the tool bar;
- Select **3D**;
- In **Insert Graph Dialogue**, choose **3D Scatter**, and click **OK**;
- In the **3D Line/Scatterplot [1] dialogue** select **Data Set** `airpoll`;
- Select *SO2* for **x Column**, *NOX* for **y Column**, and *Mortality* for **z**;
- Click **OK**.

Mortality appears to increase rapidly with increasing *NOX* values but more modestly with increasing levels of *SO2*. (A similar diagram can be found using the `cloud` function in S-PLUS and in R where it is available in the *lattice* library.)

Often it is easier to see what is happening in such a plot if lines are used to join drop-line plot. Such a plot is obtained using the instructions above but in **Insert Graph dialogue**, choose **3D Scatter with Drop Line**. The result is shown in Figure 2.16

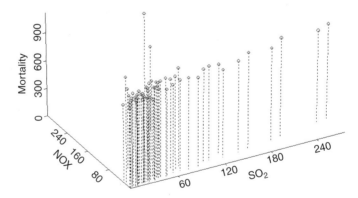

Figure 2.16 Three-dimensional drop line plot of *SO2*, *NOX*, and *Mortality*.

2.7 Conditioning Plots and Trellis Graphics

The conditioning plot or *coplot* is a potentially powerful visualization tool for studying the bivariate relationship of a pair of variables conditional on the values of one or more other variables. Such plots can often highlight the presence of interactions between the variables where the degree and/or direction of the bivariate relationship differs in the different levels of the third variable.

To illustrate we will construct a coplot of *Mortality* against *SO2* conditioned on population density (*Popden*) for the air pollution data. We need the R and S-PLUS function coplot

```
coplot(Mortality~SO2|Popden)
```

The resulting plot is shown in Figure 2.17. In this diagram, the panel at the top is known as the given panel; the panels below are dependence panels. Each rectangle in the given panel specifies a range of values of population density. On a corresponding dependence panel, *Mortality* is plotted against *SO2* for those regions with population densities within one of the intervals in the given panel. To match the latter to the dependence panels, these panels need to be examined from left to right in the bottom row and then again left to right in subsequent rows.

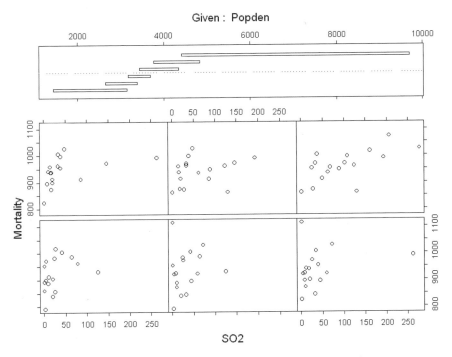

Figure 2.17 Coplot of *SO2* and *Mortality* conditional on population density.

There are relatively few observations in each panel on which to draw conclusions about possible differences in the relationship of *SO2* and *Mortality* at different levels of population density although there do appear to be some differences. In such cases it is often helpful to enhance the coplot dependence panels in some way. Here we add a locally weighted regression fit using the R and S-PLUS code:

```
coplot(Mortality~SO2|Popden,panel=function(x,y,col,pch)
    panel.smooth(x,y,span=1))
```

The result is shown in Figure 2.18. This plot suggests that the relationship between mortality and sulphur dioxide for lower levels of population density is more complex than at higher levels, although the number of points on which this claim is made is rather small.

Conditional graphical displays are simple examples of a more general scheme known as *trellis graphics* (Becker and Cleveland, 1994). This is an approach to examining high-dimensional structure in data by means of one-, two-, and three-dimensional graphs. The problem addressed is how observations of one or more variables depend on the observations of the other variables. The essential feature of this approach is the multiple conditioning that allows some type of plot to be displayed for different values of a given variable (or variables). The aim is to help in

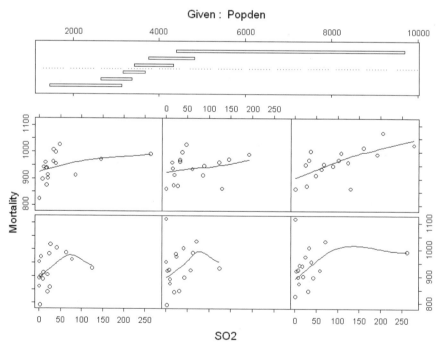

Figure 2.18 Coplot of *SO2* and *Mortality* conditional on population density with added locally weighted regression fit.

understanding both the structure of the data and how well proposed models describe
the structure. An excellent recent example of the application of trellis graphics is
given in Verbyla et al. (1999). To illustrate the possibilities we shall construct a
three-dimensional plot of *SO2*, *NOX*, and *Mortality* conditional on *Popden*. The
necessary "click-and-point" steps are:

- Click on **Data** in the tool bar;
- In **Select Data** box choose `airpoll`;
- Click **OK**;
- Click on 3D plots button, ![icon], to get 3D plot palette;
- Highlight *NOX* in spreadsheet and then ctrl click on *SO2*, *Mortality*, and *Popden*;
- Turn conditioning button ![icon] on;
- Choose **Drop line scatter** from 3D palette.

The resulting diagram is shown in Figure 2.19.

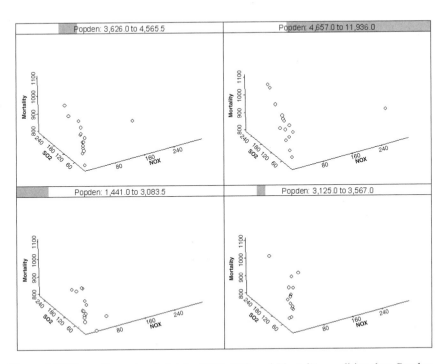

Figure 2.19 Three-dimensional plot for *NOX*, *SO2*, and *Mortality* conditional on *Popden*.

2.8 Summary

Plotting multivariate data is an essential first step in trying to understand their message. The possibilities are almost limitless with software such as R and S-PLUS and readers are encouraged to explore more fully what is available. The methods covered in this chapter provide just some basic ideas for taking an initial look at multivariate data.

Exercises

2.1   The bubbleplot makes it possible to accommodate three variable values on a scatterplot. More than three variables can be accommodated by using what might be termed a *star plot* in which the extra variables are represented by the lengths of the sides of a "star." Construct such a plot for all seven variables in the air pollution data using say *Rainfall* and *SO2* to form the basic scatterplot. (Use the `symbols` function.)

2.2   Construct a scatterplot matrix of the air pollution data in which each panel shows a bivariate density estimate of the pair of variables.

2.3   Construct a trellis graphic showing a scatterplot of *SO2* and *Mortality* conditioned on both rainfall and population density.

2.4   Construct a three-dimensional plot of *Rainfall, SO2*, and *Mortality* showing the estimated regression surface of *Mortality* on the other two variables.

2.5   Construct a three-dimensional plot of *SO2, NOX*, and *Rainfall* in which the observations are labelled by an abbreviated form of the region name.

2.6   Investigate the use of the `chiplot` function on all pairs of variables in the air pollution data.

2.7   Investigate the use of the `bkde2D` function in the *KernSmooth* library of R to calculate the bivariate density of *SO2* and *Mortality* in the air pollution data. Use the `wireframe` function available in the *lattice* library in R to construct a perspective plot of the estimated density.

2.8   Produce a similar diagram to that given in Figure 2.19 using the `cloud` function.

3
Principal Components Analysis

3.1 Introduction

The basic aim of principal components analysis is to describe the variation in a set of correlated variables, $x_1, x_2, \ldots, x_q$, in terms of a new set of uncorrelated variables, $y_1, y_2, \ldots, y_q$, each of which is a linear combination of the x variables. The new variables are derived in decreasing order of "importance" in the sense that y_1 accounts for as much of the variation in the original data amongst all linear combinations of $x_1, x_2, \ldots, x_q$. Then y_2 is chosen to account for as much as possible of the remaining variation, subject to being uncorrelated with y_1, and so on. The new variables defined by this process, $y_1, y_2, \ldots, y_q$, are the principal components.

The general hope of principal components analysis is that the first few components will account for a substantial proportion of the variation in the original variables, $x_1, x_2, \ldots, x_q$, and can, consequently, be used to provide a convenient lower-dimensional summary of these variables that might prove useful for a variety of reasons. Consider, for example, a set of data consisting of examination scores for several different subjects for each of a number of students. One question of interest might be how best to construct an informative index of overall examination performance. One obvious possibility would be the mean score for each student, although if the possible or observed range of examination scores varied from subject to subject, it might be more sensible to weight the scores in some way before calculating the average, or alternatively standardize the results for the separate examinations before attempting to combine them. In this way it might be possible to spread the students out further and so obtain a better ranking. The same result could often be achieved by applying principal components to the observed examination results and using the student's scores on the first principal component to provide a measure of examination success that maximally discriminated between them.

A further possible application for principal components analysis arises in the field of economics, where complex data are often summarized by some kind of index number, for example, indices of prices, wage rates, cost of living, and so on. When assessing changes in prices over time, the economist will wish to allow for the fact that prices of some commodities are more variable than others, or that the prices of some of the commodities are considered more important than others; in each case the index will need to be weighted accordingly. In such examples, the first principal component can often satisfy the investigators requirements.

But it is not always the first principal component that is of most interest to a researcher. A taxonomist, for example, when investigating variation in morphological measurements on animals for which all the pairwise correlations are likely to be positive, will often be more concerned with the second and subsequent components since these might provide a convenient description of aspects of an animal's "shape"; the latter will often be of more interest to the researcher than aspects of an animal's "size" which here, because of the positive correlations, will be reflected in the first principal component. For essentially the same reasons, the first principal component derived from say clinical psychiatric scores on patients may only provide an index of the severity of symptoms, and it is the remaining components that will give the psychiatrist important information about the "pattern" of symptoms.

In some applications, the principal components may be an end in themselves and might be amenable to interpretation in a similar fashion as the factors in an *exploratory factor analysis* (see Chapter 4). More often they are obtained for use as a means of constructing an informative graphical representation of the data (see later in the chapter), or as input to some other analysis. One example of the latter is provided by regression analysis. Principal components may be useful here when:

- There are too many explanatory variables relative to the number of observations.
- The explanatory variables are highly correlated.

Both situations lead to problems when applying regression techniques, problems that may be overcome by replacing the original explanatory variables with the first few principal component variables derived from them. An example will be given later and other applications of the technique are described in Rencher (1995).

A further example of when the results from a principal components analysis may be useful in the application of *multivariate analysis of variance* (see Chapter 7) is when there are too many original variables to ensure that the technique can be used with reasonable power. In such cases the first few principal components might be used to provide a smaller number of variables for analysis.

3.2 Algebraic Basics of Principal Components

The first principal component of the observations is that linear combination of the original variables whose sample variance is greatest amongst all possible such linear combinations. The second principal component is defined as that linear combination of the original variables that accounts for a maximal proportion of the remaining variance subject to being uncorrelated with the first principal component. Subsequent components are defined similarly. The question now arises as to how the coefficients specifying the linear combinations of the original variables defining each component are found. The algebra of *sample* principal components is summarized in Display 3.1.

Display 3.1
Algebraic Basis of Principal Components Analysis

- The first principal component of the observations, y_1, is the linear combination

$$y_1 = a_{11}x_1 + a_{12}x_2 + \cdots + a_{1q}x_q$$

whose sample variance is greatest among all such linear combinations.
- Since the variance of y_1 could be increased without limit simply by increasing the coefficients $a_{11}, a_{12}, \ldots, a_{1q}$ (which we will write as the vector $\mathbf{a}_1$), a restriction must be placed on these coefficients. As we shall see later, a sensible constraint is to require that the sum of squares of the coefficients, $\mathbf{a}_1'\mathbf{a}_1$ should take the value one, although other constraints are possible.
- The second principal component y_2 is the linear combination

$$y_2 = a_{21}x_1 + a_{22}x_2 + \cdots + a_{2q}x_q$$

i.e., $y_2 = \mathbf{a}_2'\mathbf{x}$ where $\mathbf{a}_2' = [a_{21}, a_{22}, \ldots, a_{2q}]$ and $\mathbf{x}' = [x_1, x_2, \ldots, x_q]$. which has the greatest variance subject to the following two conditions:

$$\mathbf{a}_2'\mathbf{a}_2 = 1,$$
$$\mathbf{a}_2'\mathbf{a}_1 = 0.$$

The second condition ensures that y_1 and y_2 are uncorrelated.
- Similarly, the jth principal component is that linear combination $y_j = \mathbf{a}_j'\mathbf{x}$ which has the greatest variance subject to the conditions

$$\mathbf{a}_j'\mathbf{a}_j = 1,$$
$$\mathbf{a}_j'\mathbf{a}_i = 0 \quad (i < j).$$

- To find the coefficients defining the first principal component we need to choose the elements of the vector $\mathbf{a}_1$ so as to maximize the variance of y_1 subject to the constraint $\mathbf{a}_1'\mathbf{a}_1 = 1$.
- To maximize a function of several variables subject to one or more constraints, the method of *Lagrange multipliers* is used. This leads to the solution that $\mathbf{a}_1$ is the eigenvector of the sample covariance matrix, $\mathbf{S}$, corresponding to its largest eigenvalue. Full details are given in Morrison (1990), and an example with $q = 2$ appears in Subsection 3.2.4.
- The other components are derived in similar fashion, with $\mathbf{a}_j$ being the eigenvector of $\mathbf{S}$ associated with its jth largest eigenvalue.
- If the eigenvalues of $\mathbf{S}$ are $\lambda_1, \lambda_2, \ldots, \lambda_q$, then since $\mathbf{a}_i'\mathbf{a}_i = 1$, the variance of the ith principal component is given by λ_i.
- The total variance of the q principal components will equal the total variance of the original variables so that

$$\sum_{i=1}^{q} \lambda_i = s_1^2 + s_2^2 + \cdots + s_q^2$$

where s_i^2 is the sample variance of x_i. We can write this more concisely as

$$\sum_{i=1}^{q} \lambda_i = \text{trace}(\mathbf{S}).$$

- Consequently, the jth principal component accounts for a proportion P_j of the total variation of the original data, where

$$P_j = \frac{\lambda_j}{\text{trace}(\mathbf{S})}.$$

- The first m principal components, where $m < q$ account for a proportion $P^{(m)}$ of the total variation in the original data, where

$$P^{(m)} = \frac{\sum_{i=1}^{m} \lambda_i}{\text{trace}(\mathbf{S})}$$

In geometrical terms it is easy to show that the first principal component defines the line of best fit (in the least squares sense) to the q-dimensional observations in the sample. These observations may therefore be represented in one dimension by taking their projection onto this line, that is, finding their first principal component score. If the observations happen to be collinear in q dimensions, this representation would account completely for the variation in the data and the sample covariance matrix would have only one nonzero eigenvalue. In practice, of course, such collinearity is extremely unlikely, and an improved representation would be given by projecting the q-dimensional observations onto the space of the best fit, this being defined by the first two principal components. Similarly, the first m components give the best fit in m dimensions. If the observations fit exactly into a space of m-dimensions, it would be indicated by the presence of $q-m$ zero eigenvalues of the covariance matrix. This would imply the presence of $q-m$ linear relationships between the variables. Such constraints are sometimes referred to as *structural relationships*.

The account of principal components given in Display 3.1 is in terms of the eigenvalues and eigenvectors of the covariance matrix, $\mathbf{S}$. In practice, however, it is far more usual to extract the components from the correlation matrix, $\mathbf{R}$. The reasons are not difficult to identify. If we imagine a set of multivariate data where the variables $x_1, x_2, \ldots, x_q$ are of completely different types, for example, length, temperature, blood pressure, anxiety rating, etc., then the structure of the principal components derived from the covariance matrix will depend on the essentially arbitrary choice of choice of units of measurement; for example, changing lengths from centimeters to inches will alter the derived components.

Additionally if there are large differences between the variances of the original variables, those whose variances are largest will tend to dominate the early components; an example illustrating this problem is given in Jolliffe (2002). Extracting the components as the eigenvectors of **R**, which is equivalent to calculating the principal components from the original variables after each has been standardized to have unit variance, overcomes these problems. It should be noted, however, that there is rarely any simple correspondence between the components derived from **S** and those derived from **R**. And choosing to work with **R** rather than with **S** involves a definite but possibly arbitrary decision to make variables "equally important."

The correlations or covariances between the original variables and the derived components are often useful in interpreting a principal components analysis. They can be obtained as shown in Display 3.2.

Display 3.2
Correlations and Covariances of Variables and Components

- The covariance of variable i with component j is given by

$$\mathrm{Cov}(x_i, y_j) = \lambda_j a_{ji}.$$

- The correlation of variable x_i with component y_j is therefore

$$r_{x_i \cdot y_j} = \frac{\lambda_j a_{ji}}{\sqrt{\mathrm{Var}(x_i)\,\mathrm{Var}(y_j)}}$$

$$= \frac{\lambda_j a_{ji}}{s_i \sqrt{\lambda_j}} = \frac{a_{ji}\sqrt{\lambda_j}}{s_i}.$$

- If the components are extracted from the correlation matrix rather than the covariance matrix, then

$$r_{x_i \cdot y_i} = a_{ji}\sqrt{\lambda_j},$$

since in this case the standard deviation, s_i, is unity.

3.2.1 Rescaling Principal Components

It is often useful to rescale principal components so that the coefficients that define them are analogous in some respects to the factor loadings in exploratory factor analysis (see Chapter 4). Again the necessary algebra is relatively simple and is outlined in Display 3.3.

Display 3.3
Rescaling Principal Components

- Let the vectors $\mathbf{a}_1, \mathbf{a}_2, \ldots, \mathbf{a}_q$, which define the principal components, be used to form a $q \times q$ matrix, $\mathbf{A} = [\mathbf{a}_1, \ldots, \mathbf{a}_q]$.
- Arrange the eigenvalues $\lambda_1, \ldots, \lambda_q$ along the main diagonal of a diagonal matrix, Λ.
- Then it can be shown that the covariance matrix of the observed variables $x_1, x_2, \ldots, x_q$ is given by

$$S = \mathbf{A}\Lambda\mathbf{A}'.$$

(We are assuming here that $\mathbf{a}_1, \mathbf{a}_2, \ldots, \mathbf{a}_q$ have been derived from $\mathbf{S}$ rather than from $\mathbf{R}$.)

- Rescaling the vectors $\mathbf{a}_1, \mathbf{a}_2, \ldots, \mathbf{a}_q$ so that the sum of squares of their elements is equal to the corresponding eigenvalue, i.e., calculating $\mathbf{a}_i^* = \lambda_i^{1/2}\mathbf{a}_i$, allows $\mathbf{S}$ to may be written more simply as

$$S = \mathbf{A}^*(\mathbf{A}^*)'$$

where $\mathbf{A}^* = [\mathbf{a}_1^*, \ldots, \mathbf{a}_q^*]$.

- In the case where components arise from a correlation matrix this rescaling leads to coefficients that are the correlations between the components and the original variables (see Display 3.2). The rescaled coefficients are analogous to factor loadings as we shall see in the next chapter. It is often these rescaled coefficients that are presented as the results of a principal components analysis.
- If the matrix $\mathbf{A}^*$ is formed from say the first m components rather than from all q, then $\mathbf{A}^*(\mathbf{A}^*)'$ gives the predicted value of $\mathbf{S}$ based on these m components.

3.2.2 *Choosing the Number of Components*

As described earlier, principal components analysis is seen to be a technique for transforming a set of observed variables into a new set of variables that are uncorrelated with one another. The variation in the original q variables is only *completely* accounted for by *all* q principal components. The usefulness of these transformed variables, however, stems from their property of accounting for the variance in decreasing proportions. The first component, for example, accounts for the maximum amount of variation possible for any linear combination of the original variables. But how useful is this artificial variation constructed from the observed variables? To answer this question we would first need to know the proportion of the total variance of the original variables for which it accounted. If, for example, 80% of the variation in a multivariate data set involving six variables could be accounted for by a simple weighted average of the variable values, then almost all the variation can be expressed along a single continuum rather than in six-dimensional space.

The principal components analysis would have provided a highly parsimonious summary (reducing the dimensionality of the data from six to one) that might be useful in later analysis.

So the question we need to ask is how many components are needed to provide an adequate summary of a given data set? A number of informal and more formal techniques are available. Here we shall concentrate on the former; examples of the use of formal inferential methods are given in Jolliffe (2002) and Rencher (1995).

The most common of the relatively ad hoc procedures that have been suggested are the following:

- Retain just enough components to explain some specified, large percentage of the total variation of the original variables. Values between 70% and 90% are usually suggested, although smaller values might be appropriate as q or n, the sample size, increases.
- Exclude those principal components whose eigenvalues are less than the average, $\sum_{i=1}^{q} \lambda_i/q$. Since $\sum_{i}^{q} \lambda_i = \mathrm{trace}(\mathbf{S})$ the average eigenvalue is also the average variance of the original variables. This method then retains those components that account for more variance than the average for the variables.
- When the components are extracted from the correlation matrix, $\mathrm{trace}(\mathbf{R}) = q$, and the average is therefore one; components with eigenvalues less than one are therefore excluded. This rule was originally suggested by Kaiser (1958), but Jolliffe (1972), on the basis of a number of simulation studies, proposed that a more appropriate procedure would be to exclude components extracted from a correlation matrix whose associated eigenvalues are less than 0.7.
- Cattell (1965) suggests examination of the plot of the λ_i against i, the so-called *scree diagram*. The number of components selected is the value of i corresponding to an "elbow" in the curve, this point being considered to be where "large" eigenvalues cease and "small" eigenvalues begin. A modification described by Jolliffe (1986) is the *log-eigenvalue diagram* consisting of a plot of $\log(\lambda_i)$ against i.

3.2.3 *Calculating Principal Component Scores*

If we decide that we need say m principal components to adequately represent our data (using one or other of the methods described in the previous subsection), then we will generally wish to calculate the scores on each of these components for each individual in our sample. If, for example, we have derived the components from the covariance matrix, $\mathbf{S}$, then the m principal component scores for individual i with original $q \times 1$ vector of variable values $\mathbf{x}_i$, are obtained as

$$y_{i1} = \mathbf{a}_1' \mathbf{x}_i$$
$$y_{i2} = \mathbf{a}_2' \mathbf{x}_i$$
$$\vdots$$
$$y_{im} = \mathbf{a}_m' \mathbf{x}_i$$

If the components are derived from the correlation matrix, then $\mathbf{x}_i$ would contain individual i's standardized scores for each variable.

The principal component scores calculated as above have variances equal to λ_j for $j = 1, \ldots, m$. Many investigators might prefer to have scores with means zero and variances equal to unity. Such scores can be found as follows:

$$\mathbf{z} = \mathbf{\Lambda}_m^{-1} \mathbf{A}_m' \mathbf{x}$$

where $\mathbf{\Lambda}_m$ is an $m \times m$ diagonal matrix with $\lambda_1, \lambda_2, \ldots, \lambda_m$ on the main diagonal, $\mathbf{A}_m = [\mathbf{a}_1, \ldots, \mathbf{a}_m]$, and $\mathbf{x}$ is the $q \times 1$ vector of standardized scores.

We should note here that the first m principal component scores are the same whether we retain all possible q components or just the first m. As we shall see in the next chapter, this is *not* the case with the calculation of factor scores.

3.2.4 *Principal Components of Bivariate Data with Correlation Coefficient r*

Before we move on to look at some practical examples of the application of principal components analysis it will be helpful to look in a little more detail at the mathematics of the method in one very simple case. This we do in Display 3.4 for bivariate data where the variables have correlation coefficient r.

Display 3.4
Principal Components of Bivariate Data with Correlation r

- Suppose we have just two variables, x_1 and x_2, measured on a sample of individuals, with sample correlation matrix given by

$$\mathbf{R} = \begin{pmatrix} 1.0 & r \\ r & 1.0 \end{pmatrix}.$$

- In order to find the principal components of the data r we need to find the eigenvalues and eigenvectors of $\mathbf{R}$.
- The eigenvalues are roots of the equation

$$|\mathbf{R} - \lambda \mathbf{I}| = 0.$$

- This leads to a quadratic equation in λ,

$$(1 - \lambda)^2 - r^2 = 0,$$

giving eigenvalues $\lambda_1 = 1 + r$, $\lambda_2 = 1 - r$. Note that the sum of the eigenvalues is 2, equal to trace $(\mathbf{R})$.
- The eigenvector corresponding to λ_1 is obtained by solving the equation

$$\mathbf{R}\mathbf{a}_1 = \lambda_1 \mathbf{a}_1$$

- This leads to the equations

$$a_{11} + ra_{12} = (1+r)a_{11}, \qquad ra_{11} + a_{12} = (1+r)a_{12}.$$

- The two equations are identical and both reduce to $a_{11} = a_{12}$.
- If we now introduce the normalization constraint, $\mathbf{a}_1'\mathbf{a}_1 = 1$ we find that

$$a_{11} = a_{12} = \frac{1}{\sqrt{2}}.$$

- Similarly, we find the second eigenvector to be given by $a_{21} = 1/\sqrt{2}$ and $a_{22} = -1/\sqrt{2}$.
- The two principal components are then given by

$$y_1 = \frac{1}{\sqrt{2}}(x_1 + x_2), \qquad y_2 = \frac{1}{\sqrt{2}}(x_1 - x_2).$$

- Notice that if $r < 0$ the order of the eigenvalues and hence of the principal components is reversed; if $r = 0$ the eigenvalues are both equal to 1 and any two solutions at right angles could be chosen to represent the two components.
- Two further points:

  1. There is an arbitrary sign in the choice of the elements of $\mathbf{a}_i$; it is customary to choose a_{i1} to be positive.
  2. The components do not depend on r, although the proportion of variance explained by each does change with r. As r tends to 1 the proportion of variance accounted for by y_1, namely $(1+r)/2$, also tends to one.

- When $r = 1$, the points all line on a straight line and the variation in the data is unidimensional.

3.3 An Example of Principal Components Analysis: Air Pollution in U.S. Cities

To illustrate a number of aspects of principal components analysis we shall apply the technique to the data shown in Table 3.1, which is again concerned with air pollution in the United States. For 41 cities in the United States the following seven variables were recorded:

SO2: Sulphur dioxide content of air in micrograms per cubic meter
Temp: Average annual temperature in °F
Manuf: Number of manufacturing enterprises employing 20 or more workers

Table 3.1 Air Pollution in U.S. Cities. From *Biometry*, 2/E, Robert R. Sokal and F. James Rohlf. Copyright © 1969, 1981 by W.H. Freeman and Company. Used with permission.

City	SO2	Temp	Manuf	Pop	Wind	Precip	Days
Phoenix	10	70.3	213	582	6.0	7.05	36
Little Rock	13	61.0	91	132	8.2	48.52	100
San Francisco	12	56.7	453	716	8.7	20.66	67
Denver	17	51.9	454	515	9.0	12.95	86
Hartford	56	49.1	412	158	9.0	43.37	127
Wilmington	36	54.0	80	80	9.0	40.25	114
Washington	29	57.3	434	757	9.3	38.89	111
Jacksonville	14	68.4	136	529	8.8	54.47	116
Miami	10	75.5	207	335	9.0	59.80	128
Atlanta	24	61.5	368	497	9.1	48.34	115
Chicago	110	50.6	3344	3369	10.4	34.44	122
Indianapolis	28	52.3	361	746	9.7	38.74	121
Des Moines	17	49.0	104	201	11.2	30.85	103
Wichita	8	56.6	125	277	12.7	30.58	82
Louisville	30	55.6	291	593	8.3	43.11	123
New Orleans	9	68.3	204	361	8.4	56.77	113
Baltimore	47	55.0	625	905	9.6	41.31	111
Detroit	35	49.9	1064	1513	10.1	30.96	129
Minneapolis	29	43.5	699	744	10.6	25.94	137
Kansas	14	54.5	381	507	10.0	37.00	99
St Louis	56	55.9	775	622	9.5	35.89	105
Omaha	14	51.5	181	347	10.9	30.18	98
Albuquerque	11	56.8	46	244	8.9	7.77	58
Albany	46	47.6	44	116	8.8	33.36	135
Buffalo	11	47.1	391	463	12.4	36.11	166
Cincinnati	23	54.0	462	453	7.1	39.04	132
Cleveland	65	49.7	1007	751	10.9	34.99	155
Columbus	26	51.5	266	540	8.6	37.01	134
Philadelphia	69	54.6	1692	1950	9.6	39.93	115
Pittsburgh	61	50.4	347	520	9.4	36.22	147
Providence	94	50.0	343	179	10.6	42.75	125
Memphis	10	61.6	337	624	9.2	49.10	105
Nashville	18	59.4	275	448	7.9	46.00	119
Dallas	9	66.2	641	844	10.9	35.94	78
Houston	10	68.9	721	1233	10.8	48.19	103

(Continued)

Table 3.1 (*Continued*)

City	SO2	Temp	Manuf	Pop	Wind	Precip	Days
Salt Lake City	28	51.0	137	176	8.7	15.17	89
Norfolk	31	59.3	96	308	10.6	44.68	116
Richmond	26	57.8	197	299	7.6	42.59	115
Seattle	29	51.1	379	531	9.4	38.79	164
Charleston	31	55.2	35	71	6.5	40.75	148
Milwaukee	16	45.7	569	717	11.8	29.07	123

Data assumed to be available as data frame usair.dat with variable names as specified in the table.

Pop: Population size (1970 census) in thousands
Wind: Average annual wind speed in miles per hour
Precip: Average annual precipitation in inches
Days: Average number of days with precipitation per year

The data were originally collected to investigate the determinants of pollution presumably by regressing *SO2* on the other six variables. Here, however, we shall examine how principal components analysis can be used to explore various aspects of the data, before looking at how such an analysis can also be used to address the determinants of pollution question.

To begin we shall ignore the *SO2* variable and concentrate on the others, two of which relate to human ecology (*Pop*, *Manuf*) and four to climate (*Temp*, *Wind*, *Precip*, *Days*). A case can be made to use negative temperature values in subsequent analyses, since then all six variables are such that high values represent a less attractive environment. This is, of course, a personal view, but as we shall see later, the simple transformation of *Temp* does aid interpretation.

Prior to undertaking a principal components analysis (or any other analysis) on a set of multivariate data, it is usually imperative to graph the data in some way so as to gain an insight into its overall structure and/or any "peculiarities" that may have an impact on the analysis. Here it is useful to construct a scatterplot matrix of the six variables, with histograms for each variable on the main diagonal. How to do this using the S-PLUS GUI (assuming the dataframe usair.dat has already been attached) has already been described in Chapter 2 (see Section 2.5). The diagram that results is shown in Figure 3.1.

A clear message from Figure 3.1 is that there is at least one city, and probably more than one, that should be considered an outlier. On the *Manuf* variable, for example, Chicago with a value of 3344 has about twice as many manufacturing enterprises employing 20 or more workers than has the city with the second highest number (Philadelphia). We shall return to this potential problem later in the chapter, but for the moment we shall carry on with a principal components analysis of the data for *all* 41 cities.

For the data in Table 3.1 it seems necessary to extract the principal components from the correlation rather than the covariance matrix, since the six variables to be

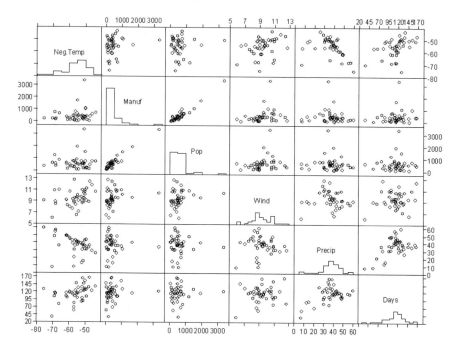

Figure 3.1 Scatterplot matrix of six variables in the air pollution data.

used are on very different scales. The correlation matrix and the principal components of the data can be obtained in R and S-PLUS® using the following command line code;

```
cor(usair.dat[,-1])
usair.pc<-princomp(usair.dat[,-1],cor=TRUE)
summary(usair.pc,loadings=TRUE)
```

The resulting output is shown in Table 3.2. (This output results from using S-PLUS; with R the signs of the coefficients of the first principal component are reversed.) One thing to note about the correlations is the very high value for *Manuf* and *Pop*, a finding returned to in Exercise 3.8. From Table 3.2 we see that the first three components all have variances (eigenvalues) greater than one and together account for almost 85% of the variance of the original variables. Scores on these three components might be used to summarize the data in further analyses with little loss of information. We shall illustrate this possibility later.

Most users of principal components analysis search for an interpretation of the derived coefficients that allow them to be "labelled" in some sense. This requires examining the coefficients defining each component (in Table 3.2 these are scaled so that their sums of squares equal unity—"blanks" indicate near-zero values), we see that the first component might be regarded as some index of "quality of life" with high values indicating a relatively poor environment (in the author's terms at least). The second component is largely concerned with a city's rainfall, having

Table 3.2 S-PLUS Results from the Principal Components Analysis of the Air Pollution Data

	Neg temp	Manuf	Pop	Wind	Precip	Days
Neg temp	1.000	0.190	0.063	0.350	−0.386	0.430
Manuf	0.190	1.000	0.955	0.238	−0.032	0.132
Pop	0.0627	0.955	1.000	0.213	−0.026	0.042
Wind	0.350	0.238	0.213	1.000	−0.013	0.164
Precip	−0.386	−0.032	−0.026	−0.013	1.000	0.496
Days	0.430	0.132	0.042	0.164	0.496	1.000

Importance of components:

	Comp 1	Comp 2	Comp 3	Comp 4	Comp 5	Comp 6
Standard deviation	1.482	1.225	1.181	0.872	0.338	0.186
Proportion of variance	0.366	0.250	0.232	0.127	0.019	0.006
Cumulative proportion	0.366	0.616 ·	0.848	0.975	0.994	1.000

Loadings:

	Comp 1	Comp 2	Comp 3	Comp 4	Comp 5	Comp 6
Neg temp	0.330	0.128	0.672	−0.306	0.558	0.136
Manuf	0.612	−0.168	−0.273	−0.137	0.102	−0.703
Pop	0.578	−0.222	−0.350	—	—	0.695
Wind	0.354	0.131	0.297	0.869	−0.113	—
Precip	—	0.623	−0.505	0.171	0.568	—
Days	0.238	0.708	—	−0.311	−0.580	—

high coefficients for *Precip* and *Days*, and might be labeled as the "wet weather" component. Component three is essentially a contrast between *Precip* and *Neg temp*, and will separate cities having high temperatures and high rainfall from those that are colder but drier. A suitable label might be simply "climate type."

Attempting to label components in this way is not without its critics; the following quotation from Marriott (1974) should act as a salutary warning about the dangers of overinterpretation.

> It must be emphasized that no mathematical method is, or could be, designed to give physically meaningful results. If a mathematical expression of this sort has an obvious physical meaning, it must be attributed to a lucky change, or to the fact that the data have a strongly marked structure that shows up in analysis. Even in the latter case, quite small sampling fluctuations can upset the interpretation; for example, the first two principal components may appear in reverse order, or may become confused altogether. Reification then requires considerable skill and experience if it is to give a true picture of the physical meaning of the data.

Even if we do not care to label the three components they can still be used as the basis of various graphical displays of the cities. In fact, this is often the most useful aspect of a principal components analysis because regarding the principal components analysis as a means to providing an informative view of multivariate data has the advantage of making it less urgent or tempting to try to interpret and label the components. The first few component scores provide a low-dimensional "map" of the observations in which the Euclidean distances between the points representing the individuals best approximate in some sense the Euclidean distances between the individuals based on the original variables. We shall return to this point in Chapter 5.

So we will begin by looking at the scatterplot of the first two principal components created using the following R and S-PLUS commands;

```
#choose square plotting area and make limits on both the x
#and y axes the same
#
par(pty="s")
plot(usair.pc$scores[,1],usair.pc$scores[,2],
ylim=range(usair.pc$scores[,1]),
xlab="PC1",ylab="PC2",type="n",lwd=2)
#
#now add abbreviated city names
#
text(usair.pc$scores[,1],usair.pc$scores[,2],
labels=abbreviate(row.names(usair.dat)),cex=0.7,lwd=2)
```

The resulting diagram is given in Figure 3.2. Similar diagrams for components 1 and 3 and 2 and 3 are given in Figures 3.3 and 3.4. (These diagrams are from the S-PLUS results.) The plots again demonstrate clearly that Chicago is an outlier and suggest that Phoenix and Philadelphia may also be suspects in this respect. Phoenix appears to offer the best quality of life (on the limited basis on the six variables recorded), and Buffalo is a city to avoid if you prefer a drier environment. We leave further interpretation to readers.

We can also construct a three-dimensional plot of the cities using these three component scores. The initial step is to construct a new data frame containing the first three principal component scores and the city names using

```
usair1.dat <- data.frame(cities=row.names(usair.dat),
    usair.dat, usair.pc$scores[,1:3])
attach(usair1.dat)
```

We shall now use the S-PLUS GUI to construct a drop-line three-dimensional plot of the data. Details of how to construct such a plot were given in Chapter 2, but it may be helpful to go through them again here;

- Click **Graph** on the tool bar;
- Select **3D**;
- In **Insert Graph** dialogue, choose **3D Scatter with drop line (x, y)**, and click **OK**;
- In the **3D Line/Scatter Plot [1] dialogue** select **Data Set** usair.dat;
- Select *Comp 1* for **x Column**, *Comp 2* for **y Column**, *Comp 3* for **z Column** and *Cities* for **w Column**;
- Check **Symbol** tab;
- Check **Use Text as Symbol** button;
- Specify text to use as **w Column**;
- Change **Font** to bold and **Height** to 0.15, click **OK**

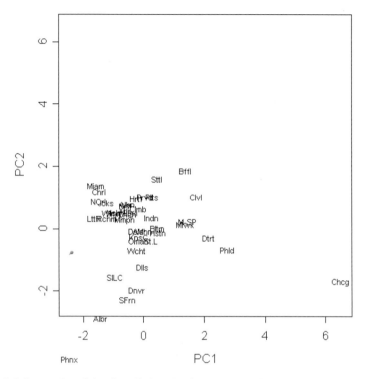

Figure 3.2 Scatterplot of the air pollution data in the space of the first two principal components.

The resulting diagram is shown in Figure 3.5. Again the problem with Chicago is very clear.

We will now use the three component scores for each city to investigate perhaps the prime question for these data, namely what characteristics of a city are predictive of its level of sulfur dioxide pollution? It may first be helpful to have a record of the component scores found from

```
usair.pc$scores[,1:3]
```

The scores are shown in Table 3.3. Before undertaking a formal regression analysis of the data we might look at *SO2* plotted against each of the three principal component scores. We can construct these plots in both R and S-PLUS as follows:

```
par(mfrow=c(1,3))
plot(usair.pc$scores[,1],SO2,xlab="PC1")
plot(usair.pc$scores[,2],SO2,xlab="PC2")
plot(usair.pc$scores[,3],SO2,xlab="PC3")
```

The plots are shown in Figure 3.6.

Interpretation of the plots is somewhat hampered by the presence of the outliers such as Chicago, but it does appear that pollution *is* related to the first principal component score but not, perhaps, to the other two. We can examine this more

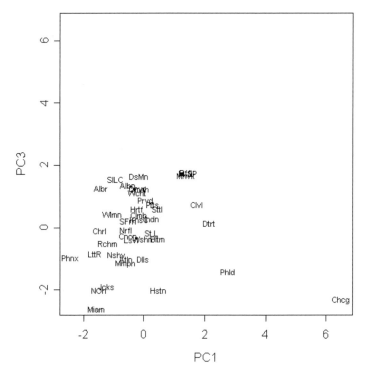

Figure 3.3 Scatterplot of the air pollution data in the space of the first and third principal components.

formally by regressing sulphur dioxide concentration on the first three principal components scores. The necessary R and S-PLUS command is;

```
summary(lm(SO2 ~ usair.pc$scores[, 1] + usair.pc$scores[, 2] +
usair.pc$scores[, 3]))
```

The resulting output is shown in Table 3.4. Clearly pollution is predicted only by the first principal component score. As "quality of life"—as measured by the human ecology and climate variable—gets worse (i.e., first PC score increases), pollution also tends to increase. (Note that because we are using principal component scores as explanatory variables in this regression the correlations of coefficients are all zero.)

Now we need to consider what to do about the obvious outliers in the data such as Chicago. The simplest approach would be to remove the relevant cities and then repeat the analyses above. The problem with such an approach is deciding when to stop removing cities, and we shall leave that as an exercise for the reader (see Exercise 3.7). Here we shall use a different approach that involves what is known as the *minimum volume ellipsoid*, a robust estimator of the correlation matrix of the data proposed by Rousseeuw (1985) and described in less technical terms in Rousseeuw and van Zomeren (1990). The essential feature of the estimator is

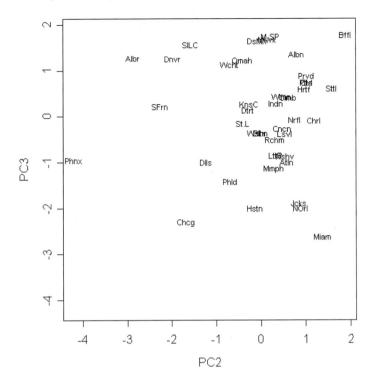

Figure 3.4 Scatterplot of the air pollution data in the space of the second and third principal components.

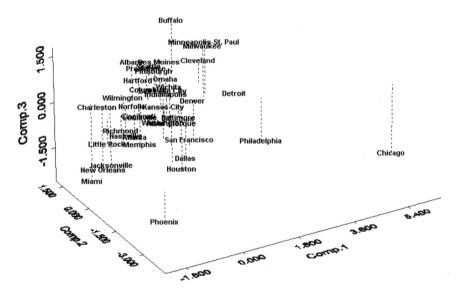

Figure 3.5 Drop line plot of air pollution data in the space of the first three principal components.

Table 3.3 First Three Principal Components Scores
for Each City in the Air Pollution Data Set

City	Comp 1	Comp 2	Comp 3
Phoenix	−2.440	−4.191	−0.942
Little Rock	−1.612	0.342	−0.840
San Francisco	−0.502	−2.255	0.227
Denver	−0.207	−1.963	1.266
Hartford	−0.219	0.976	0.595
Wilmington	−0.996	0.501	0.433
Washington	−0.023	−0.055	−0.354
Jacksonville	−1.228	0.849	−1.876
Miami	−1.533	1.405	−2.607
Atlanta	−0.599	0.587	−0.995
Chicago	6.514	−1.668	−2.286
Indianapolis	0.308	0.360	0.285
Des Moines	−0.132	−0.061	1.650
Wichita	−0.197	−0.676	1.131
Louisville	−0.424	0.541	−0.374
New Orleans	−1.454	0.901	−1.992
Baltimore	0.509	0.029	−0.364
Detroit	2.167	−0.271	0.147
Minneapolis	1.500	0.247	1.751
Kansas	−0.131	−0.252	0.275
St Louis	0.286	−0.384	−0.156
Omaha	−0.134	−0.385	1.236
Albuquerque	−1.417	−2.866	1.275
Albany	−0.539	0.792	1.363
Buffalo	1.391	1.880	1.776
Cincinnati	−0.508	0.486	−0.266
Cleveland	1.766	1.039	0.747
Columbus	−0.119	0.640	0.423
Philadelphia	2.797	−0.658	−1.415
Pittsburgh	0.322	1.027	0.748
Providence	0.070	10.34	0.888
Memphis	−0.578	0.325	−1.115
Nashville	−0.910	0.543	−0.859
Dallas	−0.007	−1.212	−0.998
Houston	0.508	−0.113	−1.994

(*Continued*)

Table 3.3 (*Continued*)

City	Comp 1	Comp 2	Comp 3
Salt Lake City	−0.912	−1.547	1.565
Norfolk	−0.589	0.752	−0.061
Richmond	−1.172	0.335	−0.509
Seattle	0.482	1.597	0.609
Charleston	−1.430	1.211	−0.079
Milwaukee	1.391	0.158	1.691

selecting a covariance matrix (**C**) and mean vector (**M**) such that the determinant of **C** is minimized subject to the number of observations for which

$$(\mathbf{x}_i - \mathbf{M})'\mathbf{C}^{-1}(\mathbf{x}_i - \mathbf{M}) \le a^2$$

is greater than or equal to h where h is the integer part of $(n + q + 1)/2$. The number a^2 is a fixed constant, usually chosen as $\chi^2_{q,0.50}$, when we expect the

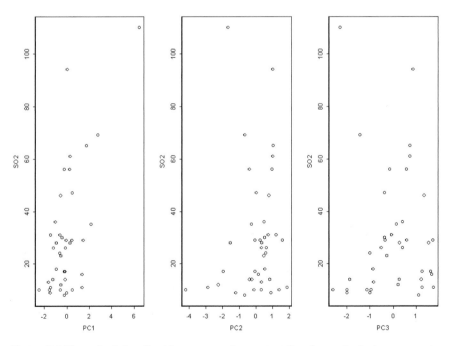

Figure 3.6 Plots of sulphur dioxide concentration against first three principal component scores.

Table 3.4 Results of Regressing Sulphur Dioxide Concentration on First Three Principal Component Scores

Residuals:

Min	1Q	Median	3Q	Max
−36.42	−10.98	−3.184	12.09	61.27

Coefficients:

	Value	Std error	t value	Pr ($?|t|$)
(Intercept)	30.0488	2.9072	10.3360	0.0000
usair.pc$scores[, 1]	9.9420	1.9617	5.0679	0.0000
usair.pc$scores[, 2]	2.2396	2.3738	0.9435	0.3516
usair.pc$scores[, 3]	−0.3750	2.4617	−0.1523	0.8798

Residual standard error: 18.62 on 37 degrees of freedom
Multiple R-squared: 0.4182
F-statistic: 8.866 on 3 and 37 degrees of freedom, the p-value is 0.0001473

Correlation of coefficients:

	(Intercept)	usair.pc$scores[, 1]	usair.pc$scores[, 2]
usair.pc$scores[, 1]	0		
usair.pc$scores[, 2]	0	0	
usair.pc$scores[, 3]	0	0	0

majority of the data to come from a normal distribution. The estimator has a high breakdown point, but is computationally expensive; see Rousseeuw and van Zomren (1990) for further details.

The necessary R and S-PLUS function to apply this estimator is `cov.mve` (in R the *lqs* library needs to be loaded to make the function available). The following code applies the function and then uses principal components on the robustly estimated correlation matrix:

```
#in R load lqs library
library(lqs)
usair.mve<-cov.mve(usair.dat[,-1],cor=TRUE)
usair.mve$cor
usair.pc1<-princomp(usair.dat[,-1],covlist=usair.mve,cor=TRUE)
summary(usair.pc1,loadings=T)
```

The resulting correlation matrix and principal components are shown in Table 3.5. (Different estimates will result each time this code is used.)

Although the pattern of correlations in Table 3.5 is largely similar to that seen in Table 3.2, there are a number of individual correlation coefficients that differ considerably in the two correlation matrices; for example, those for *Precip* and *Neg temp* (−0.386 in Table 3.2 and −0.898 in Table 3.5), and *Wind* and *Precip* (−0.013 in Table 3.2 and −0.475 in Table 3.5). The effect on the principal components analysis of these differences is, however, considerable. The first component now has a considerable negative coefficient for *Precip* and the second component is considerably different from that in Table 3.2. Labelling the coefficients is not straightfoward (at least for the author) but again it might be of interest to regress sulphur dioxide

Table 3.5 Correlation Matrix and Principal Components from Using a Robust Estimator

	Neg temp	Manuf	Pop	Wind	Precip	Days
Neg temp	1.000	0.247	0.034	0.339	−0.898	0.393
Manuf	0.247	1.000	0.842	0.292	−0.310	0.213
Pop	0.034	0.842	1.000	0.243	−0.151	0.049
Wind	0.339	0.292	0.243	1.000	−0.475	−0.109
Precip	−0.898	−0.310	−0.151	−0.475	1.000	−0.138
Days	0.393	0.213	0.049	−0.109	−0.138	1.000

Importance of components:

	Comp 1	Comp 2	Comp 3	Comp 4	Comp 5	Comp 6
Standard deviation	1.620	1.240	1.066	0.719	0.360	0.234
Proportion of variance	0.437	0.256	0.189	0.086	0.216	0.009
Cumulative proportion	0.437	0.694	0.883	0.969	0.991	1.000

Loadings:

	Comp 1	Comp 2	Comp 3	Comp 4	Comp 5	Comp 6
Neg temp	0.485	−0.455	—	−0.226	—	0.706
Manuf	0.447	0.495	−0.151	—	0.723	—
Pop	0.351	0.627	—	−0.137	−0.670	0.113
Wind	0.370	—	0.561	0.739	—	—
Precip	−0.512	0.354	−0.164	0.347	0.119	0.671
Days	0.205	−0.168	−0.791	0.504	−0.122	−0.188

concentration on the first two or three principal component scores of this second analysis; see Exercise 3.8.

3.4 Summary

Principal components analysis is among the oldest of multivariate techniques having been introduced originally by Pearson (1901) and independently by Hotelling (1933). It remains, however, one of the most widely employed methods of multivariate analysis, useful both for providing a convenient method of displaying multivariate data in a lower-dimensional space and for possibly simplifying other analyses of the data. Modern competitors to principal components analysis that may offer more powerful analyses of complex multivariate data are *projection pursuit* (Jones and Sibson, 1987), and *independent components analysis* (Hyvarinen et al., 2001). The former is a technique for finding "interesting" directions in multidimensional data sets; a brief account of the method is given in Everitt and Dunn (2001). The later is a statistical and computational technique for revealing hidden factors that underlie sets of random variables, measurements, or signals. An R implementation of both is described on the Internet at

http://CRAN.R-project.org/

Exercises

3.1 Suppose that $\mathbf{x}' = [x_1, x_2]$ is such that $x_2 = 1 - x_1$ and $x_1 = 1$ with probability p and $x_1 = 0$ with probability $q = 1 - p$. Find the covariance matrix of $\mathbf{x}$ and its eigenvalues and eigenvectors.

3.2 The eigenvectors of a covariance matrix, $\mathbf{S}$, scaled so that their sums of squares are equal to the corresponding eigenvalue, are $\mathbf{c}_1, \mathbf{c}_2, \ldots, \mathbf{c}_p$. Show that

$$\mathbf{S} = \mathbf{c}_1\mathbf{c}_1' + \mathbf{c}_2\mathbf{c}_2' + \cdots + \mathbf{c}_p\mathbf{c}_p'.$$

3.3 If the eigenvalues of $\mathbf{S}$ are $\lambda_1, \lambda_2, \ldots, \lambda_p$ show that if the coefficients defining the principal components are scaled so that $\mathbf{a}_i'\mathbf{a}_i = 1$, then the variance of the ith principal component is λ_i.

3.4 If two variables, X and Y, have covariance matrix $\mathbf{S}$ given by

$$\mathbf{S} = \begin{pmatrix} a & b \\ c & d \end{pmatrix},$$

show that if $c \neq 0$ then the first principal component is

$$\sqrt{\frac{c^2}{c^2 + (V_1 - a)^2}}X + \frac{c}{|c|}\sqrt{\frac{(V_1 - a)^2}{c^2 + (V_1 - a)^2}}Y,$$

where V_1 is the variance explained by the first principal component. What is the value of V_1?

3.5 Use S-PLUS or R to find the principal components of the following correlation matrix calculated from measurements of seven physical characteristics in each of 3000 convicted criminals:

$$R = \begin{array}{c} 1 \\ 2 \\ 3 \\ 4 \\ 5 \\ 6 \\ 7 \end{array}\begin{pmatrix} 1.00 \\ 0.402 & 1.00 \\ 0.396 & 0.618 & 1.00 \\ 0.301 & 0.150 & 0.321 & 1.00 \\ 0.305 & 0.135 & 0.289 & 0.846 & 1.00 \\ 0.339 & 0.206 & 0.363 & 0.759 & 0.797 & 1.00 \\ 0.340 & 0.183 & 0.345 & 0.661 & 0.800 & 0.736 & 1.00 \end{pmatrix}.$$

Variables:

1. Head length
2. Head breadth
3. Face breadth
4. Left finger length
5. Left forearm length
6. Left foot length
7. Height

How would you interpret the derived components?

3.6   The data in Table 3.6 show the nutritional content of different foodstuffs (the quantity involved is always three ounces). Use S-PLUS or R to create a scatterplot matrix of the data labeling the foodstuffs appropriately in each panel. On the basis of this diagram undertake what you think is an appropriate principal components analysis and try to interpret your results.

3.7   As described in the text, the air pollution data in Table 3.1 suffers from containing one or perhaps more than one outlier. Investigate this potential problem in more detail and try to reach a conclusion as to how many cities' observations

Table 3.6 Contents of Foodstuffs. From *Clustering Algorithms*, Hartigan, J.A., 1975, John Wiley & Sons, Inc. Reprinted with kind permission of J.A. Hartigan.

	Energy	Protein	Fat	Calcium	Iron
BB Beef, braised	340	20	28	9	2.6
HR Hamburger	245	21	17	9	2.7
BR Beef roast	420	15	39	7	2.0
BS Beef, steak	375	19	32	9	2.5
BC Beef, canned	180	22	10	17	3.7
CB Chicken, broiled	115	20	3	8	1.4
CC Chicken, canned	170	25	7	12	1.5
BH Beef, heart	160	26	5	14	5.9
LL Lamb leg, roast	265	20	20	9	2.6
LS Lamb shoulder, roast	300	18	25	9	2.3
HS Smoked ham	340	20	28	9	2.5
PR Pork roast	340	19	29	9	2.5
PS Pork simmered	355	19	30	9	2.4
BT Beef tongue	205	18	14	7	2.5
VC Veal cutlet	185	23	9	9	2.7
FB Bluefish, baked	135	22	4	25	0.6
AR Clams, raw	70	11	1	82	6.0
AC Clams, canned	45	7	1	74	5.4
TC Crabmeat, canned	90	14	2	38	0.8
HF Haddock, fried	135	16	5	15	0.5
MB Mackerel, broiled	200	19	13	5	1.0
MC Mackerel, canned	155	16	9	157	1.8
PF Perch, fried	195	16	11	14	1.3
SC Salmon, canned	120	17	5	159	0.7
DC Sardines, canned	180	22	9	367	2.5
UC Tuna, canned	170	25	7	7	1.2
RC Shrimp, canned	110	23	1	98	2.6

might need to be dropped before applying principal components analysis. Then undertake the analysis on the reduced data set and compare the results from those given in the text derived from using a robust estimate of the correlation matrix.

3.8 Investigate the use of the principal component scores associated with the analysis using the robust estimator of the correlation matrix as explanatory variables in a regression with sulphur dioxide concentration as dependent variable. Compare the results both with those given in Table 3.4 and those obtained in Exercise 3.7.

3.9 Investigate the use of multiple regression on the air pollution data using the human ecology and climate variables to predict sulphur dioxide pollution, keeping in mind the possible problem of the large correlation between at least two of the predictors. Do the conclusions match up to those given in the text from using principal component scores as explanatory variables?

4
Exploratory Factor Analysis

4.1 Introduction

In many areas of psychology and other disciplines in the behavioural sciences, it is often not possible to measure directly the concepts of primary interest. Two obvious examples are *intelligence* and *social class*. In such cases the researcher is forced to examine the concepts *indirectly* by collecting information on variables that *can* be measured or observed directly, and which can also realistically be assumed to be indicators, in some sense, of the concepts of real interest. The psychologist who is interested in an individual's "intelligence," for example, may record examination scores in a variety of different subjects in the expectation that these scores are related in some way to what is widely regarded as "intelligence." And a sociologist, say, concerned with people's "social class," might pose questions about a person's occupation, educational background, home ownership, etc., on the assumption that these do reflect the concept he or she is really interested in.

Both "intelligence" and "social class" are what are generally referred to as *latent variables*; i.e., concepts that cannot be measured directly but can be assumed to relate to a number of measurable or *manifest* variables. The method of analysis most generally used to help uncover the relationships between the assumed latent variables and the manifest variables is *exploratory factor analysis*. The model on which the method is based is essentially that of multiple regression, except now the manifest variables are regressed on the unobservable latent variables (often referred to in this context as *common factors*), so that direct estimation of the corresponding regression coefficients (*factor loadings*) is not possible.

4.2 The Factor Analysis Model

The basis of factor analysis is a regression model linking the manifest variables to a set of unobserved (and unobservable) latent variables. In essence the model assumes that the observed relationships between the manifest variables (as measured by their covariances or correlations) are a result of the relationships of these variables to the latent variables.

(Since it is the covariances or correlations of the manifest variables that that are central to factor analysis we can, in the description of the mathematics of the method given in Display 4.1, assume that the manifest variables all have zero mean.)

Display 4.1
Mathematics of the Factor Analysis Model

- We assume that we have a set of observed or manifest variables, $\mathbf{x}' = [x_1, x_2, \ldots, x_q]$, assumed to be linked to a smaller number of unobserved latent variables, $f_1, f_2, \ldots, f_k$ where $k < q$, by a regression model of the form

$$x_1 = \lambda_{11} f_1 + \lambda_{12} f_2 + \cdots + \lambda_{1k} f_k + u_1,$$
$$x_2 = \lambda_{21} f_1 + \lambda_{22} f_2 + \cdots + \lambda_{2k} f_k + u_2,$$
$$\vdots$$
$$x_q = \lambda_{q1} f_1 + \lambda_{q2} f_2 + \cdots + \lambda_{qk} f_k + u_q.$$

- The λ_{ij}'s are weights showing how each x_i depends on the common factors.
- The λ_{ij}'s are used in the interpretation of the factors, i.e., larger values relate a factor to the corresponding observed variables and from these we infer a meaningful description of each factor.
- The equations above may be written more concisely as

$$\mathbf{x} = \Lambda \mathbf{f} + \mathbf{u},$$

where

$$\Lambda = \begin{pmatrix} \lambda_{11} & L & \lambda_{1k} \\ \vdots & \vdots & \vdots \\ \lambda_{q1} & L & \lambda_{qk} \end{pmatrix}, \quad \mathbf{f} = \begin{pmatrix} f_1 \\ \vdots \\ f_k \end{pmatrix}, \quad \mathbf{u} = \begin{pmatrix} u_1 \\ \vdots \\ u_q \end{pmatrix}.$$

- We assume that the "residual" terms $u_1, \ldots, u_q$ are uncorrelated with each other and with the factors $f_1, \ldots, f_k$. The elements of u are specific to each x_i and hence are known as *specific variates*.
- The two assumptions above imply that, given the values of the factors, the manifest variables are independent, that is, the correlations of the observed variables arise from their relationships with the factors. In factor analysis the regression coefficients in Λ are more usually known as *factor loadings*.
- Since the factors are unobserved we can fix their location and scale arbitrarily. We shall assume they occur in standardized form with mean zero and standard deviation one. We shall also assume, initially at least, that the factors are uncorrelated with one another, in which case the factor loadings are the correlations of the manifest variables and the factors.

- With these additional assumptions about the factors, the factor analysis model implies that the variance of variable x_i, σ_i^2, is given by

$$\sigma_i^2 = \sum_{j=1}^{k} \lambda_{ij}^2 + \psi_i,$$

where ψ_i is the variance of u_i.
- So the factor analysis model implies that the variance of each observed variable can be split into two parts. The first, h_i^2, given by

$$h_i^2 = \sum_{j=1}^{k} \lambda_{ij}^2,$$

is known as the *communality* of the variable and represents the variance shared with the other variables via the common factors. The second part, ψ_i, is called the *specific* or *unique* variance, and relates to the variability in x_i not shared with other variables.
- In addition, the factor model leads to the following expression for the covariance of variables x_i and x_j:

$$\sigma_{ij} = \sum_{l=1}^{k} \lambda_{il}\lambda_{jl}.$$

- The covariances are not dependent on the specific variates in any way; the common factors above account for the relationships between the manifest variables.
- So the factor analysis model implies that the population covariance matrix, $\boldsymbol{\Sigma}$, of the observed variables has the form

$$\boldsymbol{\Sigma} = \boldsymbol{\Lambda}\boldsymbol{\Lambda}' + \boldsymbol{\Psi},$$

where

$$\boldsymbol{\Psi} = \mathrm{diag}(\psi_i).$$

- The converse also holds: If $\boldsymbol{\Sigma}$ can be decomposed into the form given above, then the k-factor model holds for $\mathbf{x}$.
- In practice, $\boldsymbol{\Sigma}$ will be estimated by the sample covariance matrix $\mathbf{S}$ (alternatively, the model will be applied to the correlation matrix $\mathbf{R}$), and we will need to obtain estimates of $\boldsymbol{\Lambda}$ and $\boldsymbol{\Psi}$ so that the observed covariance matrix takes the form required by the model (see later in the chapter for an account of estimation methods).
- We will also need to determine the value of k, the number of factors, so that the model provides an adequate fit to $\mathbf{S}$ or $\mathbf{R}$.

To apply the factor analysis model outlined in Display 4.1 to a sample of multivariate observations we need to estimate the parameters of the model in some way. The estimation problem in factor analysis is essentially that of finding $\hat{\mathbf{\Lambda}}$ and $\hat{\mathbf{\Psi}}$ for which

$$\mathbf{S} \approx \hat{\mathbf{\Lambda}}\hat{\mathbf{\Lambda}}' + \hat{\mathbf{\Psi}}.$$

(If the x_is are standardized, then $\mathbf{S}$ is replaced by $\mathbf{R}$.)

There are two main methods of estimation leading to what are known as *principal factor analysis* and *maximum likelihood factor analysis*, both of which are now briefly described.

4.2.1 *Principal Factor Analysis*

Principal factor analysis is an eigenvalue and eigenvector technique similar in many respects to principal components analysis (see Chapter 3), but operating not directly on $\mathbf{S}$ (or $\mathbf{R}$), but on what is known as the *reduced covariance matrix*, $\mathbf{S}^*$, defined as

$$\mathbf{S}^* = \mathbf{S} - \hat{\mathbf{\Psi}},$$

where $\hat{\mathbf{\Psi}}$ is a diagonal matrix containing estimates of the ψ_i.

The diagonal elements of $\mathbf{S}^*$ contain estimated *communalities*—the parts of the variance of each observed variable that can be explained by the common factors. Unlike principal components analysis, factor analysis does not try to account for *all* observed variance only that shared through the common factors. Of more concern in factor analysis is to account for the covariances or correlations between the manifest variables.

To calculate $\mathbf{S}^*$ (or with $\mathbf{R}$ replacing $\mathbf{S}$, $\mathbf{R}^*$) we need values for the communalities. Clearly we cannot calculate them on the basis of factor loadings as described in Display 4.1 since these loadings still have to be estimated. To get round this seemingly "chicken and egg" situation we need to find a sensible way of finding initial values for the communalities that does not depend on knowing the factor loadings. When the factor analysis is based on the correlation matrix of the manifest variables two frequently used methods are the following:

- Take the communality of a variable x_i as the square of the multiple correlation coefficient of x_i with the other observed variables.
- Take the communality of x_i as the largest of the absolute values of the correlation coefficients between x_i and one of the other variables.

Each of these possibilities will lead to higher values for the initial communality when x_i is highly correlated with at least some of the other manifest variables, which is essentially what is required.

Given initial communality values, a principal components analysis is performed on $\mathbf{S}^*$, and the first k eigenvectors used to provide the estimates of the loadings in the k-factor model. The estimation process can stop here or the loadings obtained at this stage ($\hat{\lambda}_{ij}$) can provide revised communality estimates calculated as $\sum_{j=1}^{k} \hat{\lambda}_{ij}^2$. The procedure is then repeated until some convergence criterion is satisfied. Difficulties can sometimes arise with this iterative approach if at any time a communality estimate exceeds the variance of the corresponding manifest variable, resulting in a negative estimate of the variable's specific variance. Such a result is known as a *Heywood case* (Heywood, 1931) and is clearly unacceptable since we cannot have a negative specific variance.

4.2.2 *Maximum Likelihood Factor Analysis*

Maximum likelihood is regarded, by statisticians at least, as perhaps the most respectable method of estimating the parameters in the factor analysis model. The essence of this approach is to define a type of "distance" measure, F, between the observed covariance matrix and the predicted value of this matrix from the factor analysis model. The measure F is defined as

$$F = \ln|\boldsymbol{\Lambda}\boldsymbol{\Lambda}' + \boldsymbol{\Psi}| + \text{trace}(\mathbf{S}|\boldsymbol{\Lambda}\boldsymbol{\Lambda}' + \boldsymbol{\Psi}|^{-1}) - \ln|\mathbf{S}| - q.$$

The function F takes the value zero if $\boldsymbol{\Lambda}\boldsymbol{\Lambda}' + \boldsymbol{\Psi}$ is equal to $\mathbf{S}$ and values greater than zero otherwise. Estimates of the loadings and the specific variances are found by minimizing F; details are given in Lawley and Maxwell (1971), Mardia et al. (1979), and Everitt (1984, 1987).

Minimizing F is equivalent to maximizing L, the likelihood function for the k-factor model, under the assumption of multivariate normality of the data, since L equals $-\frac{1}{2}nF$ plus a function of the observations. As with iterated principal factor analysis, the maximum likelihood approach can also experience difficulties with Heywood cases.

4.3 Estimating the Numbers of Factors

The decision over how many factors, k, are needed to give an adequate representation of the observed covariances or correlations is generally critical when fitting an exploratory factor analysis model. A k and $k + 1$ solution will often produce quite different factors and factor loadings for all factors, unlike a principal component analysis in which the first k components will be identical in each solution. And as pointed out by Jolliffe (1989), with too few factors there will be too many high loadings, and with too many factors, factors may be fragmented and difficult to interpret convincingly.

Choosing k might be done by examining solutions corresponding to different values of k and deciding subjectively which can be given the most convincing interpretation. Another possibility is to use the scree diagram approach described in Chapter 3, although the usefulness of this rule is not so clear in factor analysis since the eigenvalues represent variances of principal components not factors.

An advantage of the maximum likelihood approach is that it has an associated formal hypothesis testing procedure for the number of factors. The test statistic is

$$U = n' \min(F),$$

where $n' = n + 1 - \frac{1}{6}(2q + 5) - \frac{2}{3}k$. If k common factors are adequate to account for the observed covariances or correlations of the manifest variables, then U has, asymptotically, a chi-squared distribution with v degrees of freedom, where

$$v = \frac{1}{2}(q - k)^2 - \frac{1}{2}(q + k).$$

In most exploratory studies k cannot be specified in advance and so a sequential procedure is used. Starting with some small value for k (usually $k = 1$), the parameters in the corresponding factor analysis model are estimated by maximum likelihood. If U is not significant the current value of k is accepted, otherwise k is increased by one and the process repeated. If at any stage the degrees of freedom of the test become zero, then either no nontrivial solution is appropriate or alternatively the factor model itself with its assumption of linearity between observed and latent variables is questionable.

4.4 A Simple Example of Factor Analysis

The estimation procedures outlined in the previous section are needed in practical applications of factor analysis where invariably there are fewer parameters in the model than there are independent elements in $\mathbf{S}$ or $\mathbf{R}$ from which these parameters are to be estimated. Consequently the fitted model represents a genuinely parsimonious description of the data. But it is of some interest to consider a simple example in which the number of parameters is equal to the number of independent elements in $\mathbf{R}$ so that an exact solution is possible.

Spearman (1904) considered a sample of children's examination marks in three subjects—Classics (x_1), French (x_2), and English (x_3)—from which he calculated the following correlation matrix for a sample of children:

$$\mathbf{R} = \begin{array}{c} \text{Classics} \\ \text{French} \\ \text{English} \end{array} \begin{pmatrix} 1.00 & & \\ 0.83 & 1.00 & \\ 0.78 & 0.67 & 1.00 \end{pmatrix}.$$

If we assume a single factor, then the appropriate factor analysis model is

$$x_1 = \lambda_1 f + u_1,$$
$$x_2 = \lambda_2 f + u_2,$$
$$x_3 = \lambda_3 f + u_3.$$

In this example the common factor, f, might be equated with intelligence or general intellectual ability, and the specific variates, u_1, u_2, u_3 will have small variances if their associated observed variable is closely related to f. Here the number of parameters in the model (6) is equal to the number of independent elements in $\mathbf{R}$, and so by equating elements of the observed correlation matrix to the corresponding values predicted by the single-factor model we will be able to find estimates of $\lambda_1, \lambda_2, \lambda_3, \psi_1, \psi_2$, and ψ_3 such that the model fits exactly. The six equations derived from the matrix equality implied by the factor analysis model, namely

$$\mathbf{R} = \begin{bmatrix} \lambda_1 \\ \lambda_2 \\ \lambda_3 \end{bmatrix} \begin{bmatrix} \lambda_1 & \lambda_2 & \lambda_3 \end{bmatrix} + \begin{bmatrix} \psi_1 & 0 & 0 \\ 0 & \psi_2 & 0 \\ 0 & 0 & \psi_3 \end{bmatrix}$$

are

$$\hat{\lambda}_1 \lambda_2 = 0.83,$$
$$\hat{\lambda}_1 \lambda_3 = 0.78,$$
$$\hat{\lambda}_1 \lambda_4 = 0.67,$$
$$\hat{\psi}_1 = 1.0 - \hat{\lambda}_1^2,$$
$$\hat{\psi}_2 = 1.0 - \hat{\lambda}_2^2,$$
$$\hat{\psi}_3 = 1.0 - \hat{\lambda}_3^2.$$

The solutions of these equations are

$$\hat{\lambda}_1 = 0.99, \quad \hat{\lambda}_2 = 0.84, \quad \hat{\lambda}_3 = 0.79,$$
$$\hat{\psi}_1 = 0.02, \quad \hat{\psi}_2 = 0.30, \quad \hat{\psi}_3 = 0.38.$$

Suppose now that the observed correlations had been

$$\mathbf{R} = \begin{matrix} \text{Classics} \\ \text{French} \\ \text{English} \end{matrix} \begin{pmatrix} 1.00 & & \\ 0.84 & 1.00 & \\ 0.60 & 0.35 & 1.00 \end{pmatrix}.$$

In this case the solution for the parameters of a single factor model is

$$\hat{\lambda}_1 = 1.2, \quad \hat{\lambda}_2 = 0.7, \quad \hat{\lambda}_3 = 0.5,$$
$$\hat{\psi}_1 = -0.44, \quad \hat{\psi}_2 = 0.51, \quad \hat{\psi}_3 = 0.75.$$

Clearly this solution is unacceptable because of the negative estimate for the first specific variance.

4.5 Factor Rotation

Until now we have ignored one problematic feature of the factor analysis model, namely that as formulated in Display 4.1, there is no unique solution for the factor loading matrix. We can see that this is so by introducing an orthogonal matrix $\mathbf{M}$ of order $k \times k$, and rewriting the basic regression equation linking the observed and latent variables as

$$\mathbf{x} = (\mathbf{\Lambda M})(\mathbf{M'f}) + \mathbf{u}.$$

This "new" model satisfies all the requirements of a k-factor model as previously outlined with new factors $\mathbf{f}^* = \mathbf{M'f}$ and the new factor loadings $\mathbf{\Lambda M}$. This model implies that the covariance matrix of the observed variables is

$$\mathbf{\Sigma} = (\mathbf{\Lambda M})(\mathbf{\Lambda M})' + \mathbf{\Psi},$$

which, since $\mathbf{MM'} = \mathbf{I}$, reduces to $\mathbf{\Sigma} = \mathbf{\Lambda \Lambda'} + \mathbf{\Psi}$ as before. Consequently factors $\mathbf{f}$ with loadings $\mathbf{\Lambda}$ and factors $\mathbf{f}^*$ with loadings $\mathbf{\Lambda M}$ are, for any orthogonal matrix $\mathbf{M}$, equivalent for explaining the covariance matrix of the observed variables. Essentially then there are an infinite number of solutions to the factor analysis model as previously formulated.

The problem is generally solved by introducing some constraints in the original model. One possibility is to require the matrix $\mathbf{G}$ given by

$$\mathbf{G} = \mathbf{\Lambda' \Psi}^{-1} \mathbf{\Lambda}$$

to be diagonal, with its element arranged in descending order of magnitude. Such a requirement sets the first factor to have maximal contribution to the common variance of the observed variables, the second has maximal contribution to this variance subject to being uncorrelated with the first, and so on (cf. principal components analysis in Chapter 3).

The constraints on the factor loadings imposed by a condition such as that given above need to be introduced to make the parameter estimates in the factor analysis model unique. These conditions lead to orthogonal factors that are arranged in descending order of importance and enable an initial factor analysis solution to be found. The properties are not, however, inherent in the factor model, and merely considering such a solution may lead to difficulties of interpretation. For example, two consequences of these properties of a factor solution are as follows:

- The factorial complexity of variables is likely to be greater than one regardless of the underlying true model; consequently variables may have substantial loadings on more than one factor.
- Except for the first factor, the remaining factors are often *bipolar*, that is, they have a mixture of positive and negative loadings.

It may be that a more interpretable solution can be achieved using the equivalent model with loadings $\mathbf{\Lambda}^* = \mathbf{\Lambda M}$ for some particular orthogonal matrix, $\mathbf{M}$. Such a process is generally known as *factor rotation*, but before we consider how to choose $\mathbf{M}$, that is, how to "rotate" the factors, we need to address the question "Is factor rotation an acceptable process?"

Certainly in the past, factor analysis has been the subject of severe criticism because of the possibility of rotating factors. Critics have suggested that this apparently allows the investigator to impose on the data whatever type of solution they are looking for. Some have even gone so far as to suggest that factor analysis has become popular in some areas precisely because it *does* enable users to impose their preconceived ideas of the structure behind the observed correlations (Blackith and Reyment, 1971). But, on the whole, such suspicions are not justified and factor rotation can be a useful procedure for simplifying an exploratory factor analysis. Factor rotation merely allows the fitted factor analysis model to be described as simply as possible; rotation does not alter the overall structure of a solution but only how the solution is described.

Rotation is a process by which a solution is made more interpretable without changing its underlying mathematical properties. Initial factor solutions with variables loading on several factors and with bipolar factors can be difficult to interpret. Interpretation is more straightforward if each variable is highly loaded on at most one factor, and if all factor loadings are either large and positive, or near zero, with few intermediate values. The variables are thus split into disjoint sets, each of which is associated with a single factor. This aim is essentially what Thurstone (1931) referred to as *simple structure*. In more detail such structure has the following properties:

- Each row or the factor-loading matrix should contain at least one zero.
- Each column of the loading matrix should contain at least k zeros.
- Every pair of columns of the loading matrix should contain several variables whose loadings vanish in one column but not in the other.
- If the number of factors is four or more, every pair of columns should contain a large number of variables with zero loadings in both columns.
- Conversely for every pair of columns of the loading matrix only a small number of variables should have nonzero loadings in both columns.

When simple structure is achieved the observed variables will fall into mutually exclusive groups whose loadings are high on single factors, perhaps moderate to low on a few factors, and of negligible size on the remaining factors.

The search for simple structure or something close to it begins after an initial factoring has determined the number of common factors necessary and the communalties of each observed variable. The factor loadings are then transformed by post multiplication by a suitably chosen orthogonal matrix. Such a transformation is equivalent to a rigid rotation of the axes of the originally identified factor space. For a two-factor model the process of rotation can be performed graphically. As an example, consider the following correlation matrix for six school subjects:

$$\mathbf{R} = \begin{array}{c} \text{French} \\ \text{English} \\ \text{History} \\ \text{Arithmetic} \\ \text{Algebra} \\ \text{Geometry} \end{array} \begin{pmatrix} 1.00 & & & & & \\ 0.44 & 1.00 & & & & \\ 0.41 & 0.35 & 1.00 & & & \\ 0.29 & 0.35 & 0.16 & 1.00 & & \\ 0.33 & 0.32 & 0.19 & 0.59 & 1.00 & \\ 0.25 & 0.33 & 0.18 & 0.47 & 0.46 & 1.00 \end{pmatrix}.$$

The initial factor loadings are plotted in Figure 4.1. By referring each variable to the new axes shown, which correspond to a rotation of the original axes through about

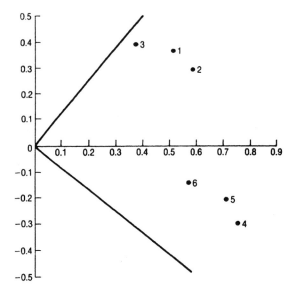

Figure 4.1 Plot of factor loadings showing a rotation of original axis.

40 degrees, a new set of loadings that give an improved description of the fitted model can be obtained. The two sets of loadings are given explicitly in Table 4.1

When there are more than two factors, more formal methods of rotation are needed. And during the rotation phase we might choose to abandon one of the assumptions made previously, namely that factors are orthogonal, that is, independent (the condition was assumed initially simply for convenience in describing the factor analysis model). Consequently two types of rotation are possible:

- *Orthogonal rotation*: methods restrict the rotated factors to being uncorrelated.
- *Oblique rotation*: methods allow correlated factors.

So the first question that needs to be considered when rotating factors is whether or not we should use an orthogonal or oblique rotation? As for many questions posed

Table 4.1 Two-Factor Solution for Correlations of Six School Subjects

Variable	Unrotated loadings		Rotated loadings	
	1	2	1	2
French	0.55	0.43	0.20	0.62
English	0.57	0.29	0.30	0.52
History	0.39	0.45	0.05	0.55
Arithmetic	0.74	−0.27	0.75	0.15
Algebra	0.72	−0.21	0.65	0.18
Geometry	0.59	−0.13	0.50	0.20

in data analysis, there is no universal answer to this question. There are advantages and disadvantages to using either type of rotation procedure. As a general rule, if a researcher is primarily concerned with getting results that "best fit" his/her data, then the researcher should rotate the factors obliquely. If, on the other hand, the researcher is more interested in the generalizability of his/her results, then orthogonal rotation is probably to be preferred.

One major advantage of an orthogonal rotation is simplicity since the loadings represent correlations between factors and manifest variables. This is *not* the case with an oblique rotation because of the correlations between the factors. Here there are two parts of the solution to consider:

- *Factor pattern coefficients*: regression coefficients that multiply with factors to produce measured variables according to the common factor model.
- *Factor structure coefficients*: correlation coefficients between manifest variables and the factors.

Additionally there is a matrix of factor correlations to consider. In many cases where these correlations are relatively small, researchers may prefer to return to an orthogonal solution.

There are a variety of rotation techniques although only relatively few are in general use. For orthogonal rotation the two most commonly used techniques are know as *varimax* and *quartimax*:

- *Varimax rotation*: originally proposed by Kaiser (1958), this has as its rationale the aim of factors with a few large loadings and as many near-zero loadings as possible. This is achieved by iterative maximization of a quadratic function of the loadings; details are given in Marda et al. (1979). This produces factors that have high correlations with one small set of variables and little or no correlation with other sets. There is a tendency for any general factor to disappear because the factor variance is redistributed.
- *Quartimax rotation*: originally suggested by Carroll (1953) this approach forces a given variable to correlate highly on one factor and either not at all or very low on other factors. Far less popular than varimax.

For oblique rotation the two methods most often used are *oblimin* and *promax*.

- *Oblimin rotation*: invented by Jennrich and Sampson (1966) this method attempts to find simple structure with regard to the factor pattern matrix through a parameter that is used to control the degree of correlation between the factors. Fixing a value for this parameter is not straightforward, but Lackey and Sullivan (2003) suggest that values between about −0.5 and 0.5 are sensible for many applications.
- *Promax rotation*: a method due to Hendrickson and White (1964) that operates by raising the loadings in an orthogonal solution (generally a varimax rotation) to some power. The goal is to obtain a solution that provides the best structure using the lowest possible power loadings and the lowest correlation between the factors.

As mentioned earlier, factor rotation is often regarded as controversial since it apparently allows the investigator to impose on the data whatever type of solution is required. But this is clearly *not* the case since although the axes may be rotated about their origin, or may be allowed to become oblique, *the distribution of the points will remain invariant*. Rotation is simply a procedure that allows new axes to be chosen so that the positions of the points can be described as simply as possible.

It should be noted that rotation techniques are also often applied to the results from a principal components analysis in the hope that it will aid in their interpretability. Although in some cases this may be acceptable, it does have several disadvantages which are listed by Jolliffe (1989). The main problem is that the defining property of principal components, namely that of accounting for maximal proportions of the total variation in the observed variables, is lost after rotation.

4.6 Estimating Factor Scores

In most applications an exploratory factor analysis will consist of the estimation of the parameters in the model and the rotation of the factors, followed by an (often heroic) attempt to interpret the fitted model. There are occasions, however, when the investigator would like to find factor scores for each individual in the sample. Such scores, like those derived in a principal components analysis (see Chapter 3), might be useful in a variety of ways. But the calculation of factor scores is not as straightforward as the calculation of principal components scores. In the original equation defining the factor analysis model, the variables are expressed in terms of the factors, whereas to calculate scores we require the relationship to be in the opposite direction. Bartholomew (1987) makes the point that to talk about "estimating" factor score is essentially misleading since they are random variables, and the issue is really one of prediction.

But if we make the assumption of normality, the conditional distribution of $\mathbf{f}$ given $\mathbf{x}$ can be found. It is

$$N[\mathbf{\Lambda}'\mathbf{\Sigma}^{-1}\mathbf{x}, (\mathbf{\Lambda}'\mathbf{\Psi}^{-1}\mathbf{\Lambda} + \mathbf{I})^{-1}].$$

Consequently, one plausible way of calculating factor scores would be to use the sample version of the mean of this distribution, namely

$$\hat{\mathbf{f}} = \hat{\mathbf{\Lambda}}'\mathbf{S}^{-1}\mathbf{x},$$

where the vector of scores for an individual, $\mathbf{x}$, is assumed to have mean zero, that is, sample means for each variable have already been subtracted. Other possible methods for deriving factor scores are described in Rencher (1995). In many respects the most damaging problem with factor analysis is not the rotational indeterminacy of the loadings, but the indeterminacy of the factor scores.

4.7 Two Examples of Exploratory Factor Analysis

4.7.1 *Expectations of Life*

The data in Table 4.2 show life expectancy in years by country, age, and sex. The data come from Keyfitz and Flieger (1971) and relate to life expectancies in the 1960s.

We will use the formal test for number of factors incorporated into the maximum likelihood approach. We can apply this test to the data, assumed to be contained in the dataframe `life` with the country names labelling the rows and variables names as given in parentheses in Table 4.2, using the following R and S-PLUS code:

```
life.fa1<-factanal(life,factors=1,method="mle")
life.fa1
life.fa2<-factanal(life,factors=2,method="mle")
life.fa2
life.fa3<-factanal(life,factors=3,method="mle")
life.fa3
```

The results from the test are shown in Table 4.3. These results indicate that a three-factor solution is adequate for the data, although it has to be remembered that with only 31 countries, use of an asymptotic test result may be rather suspect. (The numerical results from R and S-PLUS® may differ a little.)

To find the details of the three-factor solution given by maximum likelihood we use the single R instruction

```
life.fa3
```

(In S-PLUS `summary(life.fa3)` is needed.)

The results, shown in Table 4.4, correspond to a varimax-rotated solution (the default for the `factanal` function). For interest we might also compare this with results from the quartimax rotation technique. The necessary S-PLUS code to find this solution is

```
life.fa3<-factanal(life,factors=3,method="mle",
   rotation="quartimax")summary(life.fa3)
```

(R does not have the quartimax option in `factanal`.) The results are shown in Table 4.5.

The first two factors from both varimax and quartimax are similar. The first factor is dominated by life expectancy at birth for both males and females and the second reflects life expectancies at older ages. The third factor from the varimax rotation has its highest loadings for the life expectancies of men aged 50 and 75.

If using S-PLUS the estimated factor scores are already available in `life.fa3$scores`. In R the scores have to be requested as follows;

```
scores<-factanal(life,factors=3,method="mle",
scores="regression")$scores
```

Table 4.2 Life Expectancies for Different Countries by Age and Sex

Age	Male 0 (m0)	25 (m25)	50 (m50)	75 (m75)	Female 0 (w0)	25 (w25)	50 (w50)	75 (w75)
Algeria	63	51	30	13	67	54	34	15
Cameroon	34	29	13	5	38	32	17	6
Madagascar	38	30	17	7	38	34	20	7
Mauritius	59	42	20	6	64	46	25	8
Reunion	56	38	18	7	62	46	25	10
Seychelles	62	44	24	7	69	50	28	14
South Africa (B)	50	39	20	7	55	43	23	8
South Africa (W)	65	44	22	7	72	50	27	9
Tunisia	56	46	24	11	63	54	33	19
Canada	69	47	24	8	75	53	29	10
Cost Rica	65	48	26	9	68	50	27	10
Dominican Republic	64	50	28	11	66	51	29	11
El Salvador	56	44	25	10	61	48	27	12
Greenland	60	44	22	6	65	45	25	9
Grenada	61	45	22	8	65	49	27	10
Guatemala	49	40	22	9	51	41	23	8
Honduras	59	42	22	6	61	43	22	7
Jamaica	63	44	23	8	67	48˙	26	9
Mexico	59	44	24	8	63	46	25	8
Nicaragua	65	48	28	14	68	51	29	13
Panama	65	48	26	9	67	49	27	10
Trinidad (62)	64	63	21	7	68	47	25	9
Trinidad (67)	64	43	21	6	68	47	24	8
United States (66)	67	45	23	8	74	51	28	10
United States (NW66)	61	40	21	10	67	46	25	11
United States (W66)	68	46	23	8	75	52	29	10
United States (67)	67	45	23	8	74	51	28	10
Argentina	65	46	24	9	71	51	28	10
Chile	59	43	23	10	66	49	27	12
Colombia	58	44	24	9	62	47	25	10
Ecuador	57	46	28	9	60	49	28	11

Table 4.3 Results from Test for Number of Factors on the Data in Table 4.2 Using R

1. Test of the hypothesis that one factor is sufficient versus the alternative that more are required:

   The chi square statistic is 163.11 on 20 degrees of freedom. The p-value is <0.0001.

2. Test of the hypothesis that two factors are sufficient versus the alternative that more are required:

   The chi square statistic is 45.24 on 13 degrees of freedom. The p-value is <0.0001.

3. Test of the hypothesis that three factors are sufficient versus the alternative that more are required:

   The chi square statistic is 6.73 on 7 degrees of freedom. The p-value is 0.458.

Table 4.4 Maximum Likelihood Three-Factor Solution for Life Expectancy Data After Varimax Rotation Using R

Importance of factors:

	Factor 1	Factor 2	Factor 3
SS loadings	3.38	2.08	1.64
Proportion Var	0.42	0.26	0.21
Cumulative Var	0.42	0.68	0.89

The degrees of freedom for the model is 7.

Uniquenesses:

M0	M25	M50	M75	W0	W25	W50	W75
0.005	0.362	0.066	0.288	0.005	0.011	0.020	0.146

Loadings:

	Factor 1	Factor 2	Factor 3	Communality
M0	0.97	0.12	0.23	0.9999
M25	0.65	0.17	0.44	0.6491
M50	0.43	0.35	0.79	0.9018
M75	—	0.53	0.66	0.7077
W0	0.97	0.22	—	0.9951
W25	0.76	0.56	0.31	0.9890
W50	0.54	0.73	0.40	0.9793
W75	0.16	0.87	0.28	0.8513

The factor scores are shown in Table 4.6 (again the scores from R and S-PLUS may differ a little). We can use the scores to provide a 3-D plot of the data by first creating a new dataframe

```
#if using S-PLUS we need scores<-life.fa3$scores
lifex<-data.frame(life,scores)
attach(lifex)
```

Table 4.5 Three-Factor Solution for Life Expectancy Data After Quartimax Rotation Using S-PLUS

	Factor 1	Factor 2	Factor 3
SS loadings	4.57	2.13	0.37
Proportion Var	0.57	0.26	0.04
Cumulative Var	0.57	0.84	0.88

The degrees of freedom for the model is 7.

Uniquenesses:

M0	M25	M50	M75	W0	W25	W50	W75
0.0000876	0.3508347	0.09818739	0.2923066	0.004925743	0.01100307	0.02074596	0.1486658

Loadings:

	Factor 1	Factor 2	Factor 3	Communality
M0	0.99	—	—	0.9999
M25	0.76	0.18	0.21	0.6491
M50	0.66	0.57	0.37	0.9018
M75	0.33	0.74	0.23	0.7077
W0	0.98	—	−0.16	0.9951
W25	0.90	0.39	−0.14	0.9890
W50	0.74	0.65	−0.14	0.9793
W75	0.37	0.80	−0.28	0.8513

Table 4.6 Factor Scores from the Three-Factor Solution for the Life Expectancy Data

	Factor 1	Factor 2	Factor 3
Algeria	−0.26	1.92	1.96
Cameroon	−2.84	−0.69	−1.98
Madagascar	−2.82	−1.03	0.29
Mauritius	0.15	−0.36	−0.77
Reunion	−0.19	0.35	−1.39
Seychelles	0.38	0.90	−0.71
South Africa (B)	−1.07	0.06	−0.87
South Africa (W)	0.95	0.12	−1.02
Tunisia	−0.87	3.52	−0.21
Canada	1.27	0.26	−0.22
Cost Rica	0.52	−0.52	1.06
Dominican Republic	0.11	−0.01	1.94
El Salvador	−0.64	0.82	0.25
Greenland	0.24	−0.67	−0.45
Grenada	0.15	0.11	0.08
Guatemala	−1.48	−0.64	0.62
Honduras	0.07	−1.93	0.38

(*Continued*)

Table 4.6 (*Continued*)

Jamaica	0.48	−0.58	0.17
Mexico	−0.07	−0.60	0.26
Nicaragua	0.28	0.08	1.77
Panama	0.47	−0.84	1.43
Trinidad (62)	0.72	−1.07	−0.00
Trinidad (67)	0.82	−1.24	−0.36
United States (66)	1.14	0.20	−0.75
United States (NW66)	0.41	−0.39	−0.74
United States (W66)	1.23	0.40	−0.68
United States (67)	1.14	0.20	−0.75
Argentina	0.73	0.31	−0.21
Chile	−0.02	0.91	−0.73
Colombia	−0.26	−0.19	0.28
Ecuador	−0.75	0.62	1.36

and then using the S-PLUS GUI as described in Chapter 2. The resulting diagram is shown in Figure 4.2.

Ordering along the first axis reflects life expectancy at birth ranging from Cameroon and Madagascar to countries such as the United States. And on the third axis Algeria is prominent because it has high life expectancy amongst men at higher ages with Cameroon at the lower end of the scale with a low life expectancy for men over 50.

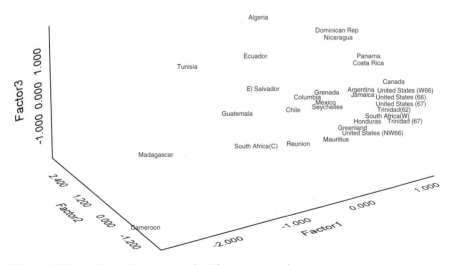

Figure 4.2 Plot of three-factor scores for life expectancy data.

4.7.2 *Drug Usage by American College Students*

The majority of adult and adolescent Americans regularly use psychoactive substances during an increasing proportion of their lifetime. Various forms of licit and illicit psychoactive substances use are prevalent, suggesting that patterns of psychoactive substance taking are a major part of the individual's behavioural repertory and have pervasive implications for the performance of other behaviors. In an investigation of these phenomena, Huba et al. (1981) collected data on drug usage rates for 1634 students in the seventh to ninth grades in 11 schools in the greater metropolitan area of Los Angeles. Each participant completed a questionnaire about the number of times a particular substance had ever been used. The substances asked about were as follows:

X1. cigarettes
X2. beer
X3. wine
X4. liquor
X5. cocaine
X6. tranquillizers
X7. drug store medications used to get high
X8. heroin and other opiates
X9. marijuana
X10. hashish
X11. inhalants (glue, gasoline, etc.)
X12. hallucinogenics (LSD, mescaline, etc.)
X13. amphetamine, stimulants

Responses were recorded on a five-point scale;

1. never tried
2. only once
3. a few times
4. many times
5. regularly

The correlations between the usage rates of the 13 substances are shown in Table 4.7. We first try to determine the number of factors using the maximum likelihood test. Here the S-PLUS code needs to accommodate the use of the correlation matrix rather than the raw data. We assume the correlation matrix is available as the data frame druguse.cor. The R code for finding the results of the test for number of factors here is

R

```
druguse.fa<-lapply(1:6,function(nf)
factanal(covmat=druguse.cor,factors=nf,method="mle",
  n.obs=1634)
```

Table 4.7 Correlation Matrix for Drug Usage Data

	X1	X2	X3	X4	X5	X6	X7	X8	X9	X10	X11	X12	X13
X1	1												
X2	0.447	1											
X3	0.442	0.619	1										
X4	0.435	0.604	0.583	1									
X5	0.114	0.068	0.053	0.115	1								
X6	0.203	0.146	0.139	0.258	0.349	1							
X7	0.091	0.103	0.110	0.122	0.209	0.221	1						
X8	0.082	0.063	0.066	0.097	0.321	0.355	0.201	1					
X9	0.513	0.445	0.365	0.482	0.186	0.315	0.150	0.154	1				
X10	0.304	0.318	0.240	0.368	0.303	0.377	0.163	0.219	0.534	1			
X11	0.245	0.203	0.183	0.255	0.272	0.323	0.310	0.288	0.301	0.302	1		
X12	0.101	0.088	0.074	0.139	0.279	0.367	0.232	0.320	0.204	0.368	0.304	1	
X13	0.245	0.199	0.184	0.293	0.278	0.545	0.232	0.314	0.394	0.467	0.392	0.511	1

The S-PLUS code here is a little different

S-PLUS
```
druguse.list<-list(cov=druguse.cor,center=rep(0,13),
  n.obs=1634)
druguse.fa<-lapply(1:6,function(nf)
  factanal(covlist=druguse.list,factors=nf,method="mle"))
```

The results from the test of number of factors are shown in Table 4.8. The test suggests that a six-factor model is needed. The results from the six-factor varimax solution are obtained from

R: `druguse.fa[[6]]`
S-PLUS: `summary(druguse.fa[[6]])`

and are shown in Table 4.9. The first factor involves cigarettes, beer, wine, liquor, and marijuana and we might label it "social/soft drug usage." The second factor has high loadings on cocaine, tranquillizers, and heroin. The obvious label for the factor is "hard drug usage." Factor three is essentially simply amphetamine use, and factor four is hashish use. We will not try to interpret the last two factors even though the formal test for number of factors indicated that a six-factor solution was necessary. It may be that we should not take the results of the formal test too literally. Rather, it may be a better strategy to consider the value of k indicated by the test to be an upper bound on the number of factors with practical importance. Certainly a six-factor solution for a data set with only 13 manifest variables might be regarded as not entirely satisfactory, and clearly we would have some difficulties interpreting all the factors.

Table 4.8 Results of Formal Test for Number of
Factors on Drug Usage Data from R

1. Test of the hypothesis that one factor is sufficient
   versus the alternative that more are required:

   The chi square statistic is 2278.25 on 65 degrees of freedom.
   The p-value is <0.00001.

2. Test of the hypothesis that two factor is sufficient
   versus the alternative that more are required:

   The chi square statistic is 477.37 on 53 degrees of freedom.
   The p-value is <0.00001.

3. Test of the hypothesis that three factors are sufficient
   versus the alternative that more are required:

   The chi square statistic is 231.95 on 42 degrees of freedom.
   The p-value is <0.00001.

4. Test of the hypothesis that four factors are sufficient
   versus the alternative that more are required:

   The chi square statistic is 113.42 on 32 degrees of freedom.
   The p-value is <0.00001.

5. Test of the hypothesis that five factors are sufficient
   versus the alternative that more are required:

   The chi square statistic is 60.57 on 23 degrees of freedom.
   The p-value is <0.00001.

6. Test of the hypothesis that six factors are sufficient
   versus the alternative that more are required:

   The chi square statistic is 23.97 on 15 degrees of freedom.
   The p-value is 0.066.

One of the problems is that with the large sample size in this example, even small discrepancies between the correlation matrix predicted by a proposed model and the observed correlation matrix may lead to rejection of the model. One way to investigate this possibility is to simply look at the differences between the observed and predicted correlations. We shall do this first for the six-factor model using the following R and S-PLUS code:

```
pred<-druguse.fa[[6]]$loadings%*%t(druguse.fa[[6]]
    $loadings)+
diag(druguse.fa[[6]]$uniquenesses)
druguse.cor-pred
```

The resulting matrix of differences is shown in Table 4.10. The differences are all very small, underlining that the six-factor model does describe the data very well.

Now let us look at the corresponding matrices for the three- and four-factor solutions found in a similar way; see Table 4.11. Again in both cases the residuals are all relatively small, suggesting perhaps that use of the formal test for number of

Table 4.9 Maximum Likelihood of Six-Factor Solution for Drug Usage Data—Varimax Rotation

	Factor 1	Factor 2	Factor 3	Factor 4	Factor 5	Factor 6
SS loadings	2.30	1.43	1.13	0.95	0.68	0.61
Proportion Var	0.18	0.11	0.09	0.07	0.05	0.05
Cumulative Var	0.80	0.29	0.37	0.45	0.50	0.55

The degrees of freedom for the model is 15.

Uniquenesses:

X1	X2	X3	X4	X5	X6	X7	X8	X9	X10	X11	X12	X13
0.560	0.368	0.374	0.411	0.681	0.526	0.748	0.665	0.324	0.025	0.597	0.630	4r-010

Loadings:

	Factor 1	Factor 2	Factor 3	Factor 4	Factor 5	Factor 6
X1	0.49	—	—	—	0.41	—
X2	0.78	—	—	0.10	0.11	—
X3	0.79	—	—	—	—	—
X4	0.72	0.12	0.10	0.12	0.16	—
X5	—	0.52	—	0.13	—	0.16
X6	0.13	0.56	0.32	0.10	0.14	—
X7	—	0.24	—	—	—	0.42
X8	—	0.54	0.10	—	—	0.19
X9	0.43	0.16	0.15	0.26	0.60	0.10
X10	0.24	0.28	0.19	0.87	0.20	—
X11	0.17	0.32	0.16	—	0.15	0.47
X12	—	0.39	0.34	0.19	—	0.26
X13	0.15	0.34	0.89	0.14	0.14	0.17

factors leads, in this case, to overfitting. The three-factor model appears to provide a perfectly adequate fit for these data.

4.8 Comparison of Factor Analysis and Principal Components Analysis

Factor analysis, like principal components analysis, is an attempt to explain a set of multivariate data using a smaller number of dimensions than one begins with, but the procedures used to achieve this goal are essentially quite different in the two approaches. Some differences between the two are as follows:

- Factor analysis tries to explain the covariances or correlations of the observed variables by means of a few common factors. Principal components analysis is primarily concerned with explaining the variance of the observed variables.
- If the number of retained components is increased, say, from m to $m + 1$, the first m components are unchanged. This is not the case in factor analysis, where there can be substantial changes in *all* factors if the number of factors is unchanged.
- The calculation of principal component scores is straightforward. The calculation of factor scores is more complex, and a variety of methods have been suggested.

Table 4.10 Differences Between Observed and Predicted Correlations for Six-Factor Model Fitted to Drug Usage Data

	X1	X2	X3	X4	X5	X6	X7	X8	X9	X10	X11	X12	X13
X1	0.000	-0.001	0.015	-0.018	0.010	0.000	-0.020	-0.005	0.002	0	0.013	-0.003	0
X2	-0.001	0.000	-0.002	0.004	0.004	-0.011	-0.002	0.007	0.002	0	-0.004	0.005	0
X3	0.015	-0.002	0.000	-0.001	0.000	-0.005	0.007	0.008	-0.004	0	-0.008	-0.001	0
X4	-0.018	0.004	-0.001	0.000	-0.07	0.020	-0.002	-0.018	0.004	0	0.013	-0.004	0
X5	0.010	0.004	0.000	-0.07	0.000	0.002	0.005	0.003	-0.004	0	-0.002	-0.008	0
X6	0.000	-0.011	-0.005	0.020	0.002	0.000	0.011	-0.004	-0.003	0	-0.002	-0.008	0
X7	-0.020	-0.002	0.007	-0.002	0.005	0.011	0.002	-0.018	0.007	0	0.005	-0.003	0
X8	-0.005	0.007	0.008	-0.018	0.003	-0.004	-0.018	0.001	0.006	0	0.002	0.021	0
X9	0.002	0.002	-0.004	0.004	-0.004	-0.003	0.007	0.006	0.000	0	-0.007	0.002	0
X10	0.000	0.000	0.000	0.000	0.000	0.000	0.000	0.000	0.000	0	0.000	0.000	0
X11	0.013	-0.004	-0.008	0.013	-0.002	-0.002	0.005	0.002	-0.007	0	-0.003	0.017	0
X12	-0.003	0.005	-0.001	-0.004	-0.008	-0.008	-0.003	0.021	0.002	0	-0.019	-0.003	0
X13	0.000	0.000	0.000	0.000	0.000	0.000	0.000	0.000	0.000	0	0.000	0.000	0

Table 4.11 Differences Between Observed and Predicted Correlations for Three- and Four-Factor Models Fitted to Drug Usage Data

1. Three factor

	X1	X2	X3	X4	X5	X6	X7	X8	X9	X10	X11	X12	X13
X1	0.000	-0.001	0.009	-0.013	0.011	0.009	-0.011	-0.004	0.002	-0.026	0.039	-0.017	0.002
X2	-0.001	0.000	-0.001	0.002	0.002	-0.014	0.000	0.005	-0.001	0.019	-0.003	0.009	-0.007
X3	0.009	-0.001	0.000	0.000	-0.002	-0.004	0.012	0.013	0.001	-0.017	-0.007	0.004	0.002
X4	-0.013	0.002	0.000	0.000	-0.008	0.023	-0.018	-0.020	-0.001	0.014	-0.002	-0.015	0.005
X5	0.011	0.002	-0.002	-0.008	0.000	0.030	0.038	0.082	-0.002	0.041	0.026	-0.027	-0.076
X6	0.009	-0.014	-0.004	0.023	0.030	0.002	-0.022	0.024	-0.001	-0.017	-0.035	-0.057	0.040
X7	-0.011	0.000	0.012	-0.018	0.038	-0.022	0.000	0.021	0.006	-0.040	0.116	0.002	-0.038
X8	-0.004	0.005	0.013	-0.020	0.082	0.024	0.021	0.000	0.006	-0.035	0.034	-0.003	-0.050
X9	0.002	-0.001	0.001	-0.001	-0.002	-0.001	0.006	0.006	0.000	0.001	0.006	-0.002	-0.002
X10	-0.026	0.019	-0.017	0.014	0.041	-0.017	-0.040	-0.035	0.001	0.000	-0.033	0.036	0.010
X11	0.039	-0.003	-0.007	-0.002	0.026	-0.035	0.116	0.034	0.006	-0.033	0.000	0.014	-0.012
X12	-0.017	0.009	0.004	-0.015	-0.027	-0.057	0.002	-0.003	-0.002	0.036	0.014	0.000	0.044
X13	0.002	-0.007	0.002	0.005	-0.076	0.040	-0.038	-0.050	-0.002	0.010	-0.012	0.044	0.001

2. Four factor

	X1	X2	X3	X4	X5	X6	X7	X8	X9	X10	X11	X12	X13
X1	0.000	-0.001	0.008	-0.012	0.010	0.008	-0.014	-0.007	0.001	-0.023	0.037	-0.020	0
X2	-0.001	0.000	-0.001	0.001	0.001	-0.016	-0.002	0.003	-0.001	0.018	-0.005	0.006	0
X3	0.008	-0.001	0.000	0.000	-0.001	-0.005	0.012	0.014	0.001	-0.020	-0.008	0.001	0
X4	-0.012	0.001	0.000	0.000	-0.005	0.029	-0.015	-0.016	-0.001	0.018	0.001	-0.009	0
X5	0.010	0.001	-0.001	-0.005	0.000	0.020	-0.014	0.004	-0.003	0.033	-0.018	-0.026	0
X6	0.008	-0.016	-0.005	0.029	0.020	0.000	-0.021	0.024	-0.001	0.001	-0.029	-0.025	0
X7	-0.014	-0.002	0.012	-0.015	-0.014	-0.021	0.000	0.024	-0.001	-0.042	0.095	0.011	0
X8	-0.007	0.003	0.014	-0.016	0.004	0.024	0.024	0.000	0.003	-0.038	0.003	0.008	0
X9	0.001	-0.001	0.001	-0.001	-0.003	-0.001	-0.001	0.003	0.000	0.000	0.000	0.000	0
X10	-0.023	0.018	-0.020	0.018	0.033	0.001	-0.042	-0.038	0.000	0.000	0.000	0.057	0
X11	0.037	-0.005	-0.008	0.001	-0.018	-0.029	0.095	0.003	0.000	0.000	0.000	-0.006	0
X12	-0.020	0.006	0.001	-0.009	-0.026	-0.025	0.011	0.008	0.000	0.057	-0.006	0.000	0
X13	0.000	0.000	0.000	0.000	0.000	0.000	0.000	0.000	0.000	0.000	0.000	0.000	0

- There is usually no relationship between the principal components of the sample correlation matrix and the sample covariance matrix. For maximum likelihood factor analysis, however, the results of analyzing either matrix are essentially equivalent (this is not true of principal factor analysis).

Despite these differences, the results from both types of analysis are frequently very similar. Certainly if the specific variances are small we would expect both forms of analysis to give similar results. However, if the specific variances are large they will be absorbed into all the principal components, both retained and rejected, whereas factor analysis makes special provision for them.

Lastly, it should be remembered that both principal components analysis and factor analysis are similar in one important respect—they are both pointless if the observed variables are almost uncorrelated. In this case factor analysis has nothing to explain and principal components analysis will simply lead to components which are similar to the original variables.

4.9 Confirmatory Factor Analysis

The methods described in this chapter have been those of exploratory factor analysis. In such models no constraints are placed on which of the manifest variables load on the common factors. But there is an alternative approach known as *confirmatory factor analysis* in which specific constraints *are* introduced, for example, that particular manifest variables are related to only one of the common factors with their loadings on other factors set a priori to be zero. These constraints may be suggested by theoretical considerations or perhaps from earlier exploratory factor analyses on similar data sets. Fitting confirmatory factor analysis models requires specialized software and readers are referred to Dunn et al. (1993) and Muthen and Muthen (1998).

4.10 Summary

Factor analysis has probably attracted more critical comments than any other statistical technique. Hills (1977), for example, has gone so far as to suggest that factor analysis is not worth the time necessary to understand it and carry it out. And Chatfield and Collins (1980) recommend that factor analysis should not be used in most practical situations. The reasons that these authors and others are so openly sceptical about factor analysis arises firstly from the central role of latent variables in the factor analysis model and secondly from the lack of uniqueness of the factor loadings in the model that gives rise to the possibility of rotating factors. It certainly is the case that since the common factors cannot be measured or observed, the existence of these hypothetical variables is open to question. A factor is a construct operationally defined by its factor loadings, and overly enthusiastic reification is not to be recommended.

It is the case that given one factor-loading matrix, there are an infinite number of factor-loading matrices that could equally well (or equally badly) account for the variances and covariances of the manifest variables. Rotation methods are designed to find an easily interpretable solution from among this infinitely large set of alternatives by finding a solution that exhibits the best simple structure.

Factor analysis can be a useful tool for investigating particular features of the structure of multivariate data. Of course, like many models used in data analysis, the one used in factor analysis may be only a very idealized approximation to the truth. Such an approximation may, however, prove a valuable starting point for further investigations.

Exercises

4.1 Show how the result $\Sigma = \Lambda\Lambda' + \Psi$ arises from the assumptions of uncorrelated factors, independence of the specific variates, and independence of common factors and specific variances. What form does Σ take if the factors are allowed to be correlated?

4.2 Show that the communalities in a factor analysis model are unaffected by the transformation $\Lambda^* = \Lambda M$.

4.3 Give a formula for the proportion of variance explained by the jth factor estimated by the principal factor approach.

4.4 Apply the factor analysis model separately to the life expectancies of men and women and compare the results.

4.5 Apply principal factor analysis to the drug usage data and compare the results with those given in the text from maximum likelihood factor analysis. Investigate the use of oblique rotation for these data.

4.6 The correlation matrix given below arises from the scores of 220 boys in six school subjects: (1) French, (2) English, (3) history, (4) arithmetic, (5) algebra, and (6) geometry. The two-factor solution from a maximum likelihood factor analysis is shown in Table 4.12. By plotting the derived loadings, find an

Table 4.12 Maximum Likelihood Factor Analysis for School Subjects Data

Subject	Factor loadings		Communality
	F1	F2	
1. French	0.55	0.43	0.49
2. English	0.57	0.29	0.41
3. History	0.39	0.45	0.36
4. Arithmetic	0.74	−0.27	0.62
5. Algebra	0.72	−0.21	0.57
6. Geometry	0.60	−0.13	0.37

orthogonal rotation that allows easier interpretation of the results.

$$
\mathbf{R} =
\begin{array}{l}
\text{French} \\
\text{English} \\
\text{History} \\
\text{Arithmetic} \\
\text{Algebra} \\
\text{Geometry}
\end{array}
\left(
\begin{array}{llllll}
1.00 & & & & & \\
0.44 & 1.00 & & & & \\
0.41 & 0.35 & 1.00 & & & \\
0.29 & 0.35 & 0.16 & 1.00 & & \\
0.33 & 0.32 & 0.19 & 0.59 & 1.00 & \\
0.25 & 0.33 & 0.18 & 0.47 & 0.46 & 1.00
\end{array}
\right).
$$

4.7   The matrix below shows the correlations between ratings on nine statements about pain made by 123 people suffering from extreme pain. Each statement was scored on a scale from 1 to 6 ranging from agreement to disagreement. The nine pain statements were as follows:

1. Whether or not I am in pain in the future depends on the skills of the doctors.
2. Whenever I am in pain, it is usually because of something I have done or not done.
3. Whether or not I am in pain depends on what the doctors do for me.
4. I cannot get any help for my pain unless I go to seek medical advice.
5. When I am in pain I know that it is because I have not been taking proper exercise or eating the right food.
6. People's pain results from their own carelessness.
7. I am directly responsible for my pain.
8. Relief from pain is chiefly controlled by the doctors.
9. People who are never in pain are just plain lucky.

$$
\mathbf{R} =
\left(
\begin{array}{lllllllll}
1.00 & & & & & & & & \\
-0.04 & 1.00 & & & & & & & \\
0.61 & -0.07 & 1.00 & & & & & & \\
0.45 & -0.12 & 0.59 & 1.00 & & & & & \\
0.03 & 0.49 & 0.03 & -0.08 & 1.00 & & & & \\
-0.29 & 0.43 & -0.13 & -0.21 & 0.47 & 1.00 & & & \\
-0.30 & 0.30 & -0.24 & -0.19 & 0.41 & 0.63 & 1.00 & & \\
0.45 & -0.31 & 0.59 & 0.63 & -0.14 & -0.13 & -0.26 & 1.00 & \\
0.30 & -0.17 & 0.32 & 0.37 & -0.24 & -0.15 & -0.29 & 0.40 & 1.00
\end{array}
\right)
$$

(a) Perform a principal components analysis on these data and examine the associated scree plot to decide on the appropriate number of components.
(b) Apply maximum likelihood factor analysis and use the test described in the chapter to select the necessary number of common factors.
(c) Rotate the factor solution selected using both an orthogonal and an oblique procedure, and interpret the results.

5
Multidimensional Scaling and Correspondence Analysis

5.1 Introduction

In Chapter 3 we noted in passing that one of the most useful ways of using principal component analysis was to obtain a low-dimensional "map" of the data that preserved as far as possible the Euclidean distances between the observations in the space of the original q variables. In this chapter we will make this aspect of principal component analysis more explicit and also introduce some other, more direct methods, which aim to produce similar maps of data that have a different form from the usual multivariate data matrix, $\mathbf{X}$. We will consider two such techniques The first, *multidimensional scaling*, is used, essentially, to represent an observed *proximity matrix* geometrically. Proximity matrices arise either directly from experiments in which subjects are asked to assess the similarity of pairs of stimuli, or indirectly; as a measure of the correlation, covariance, or distance of the pair of stimuli derived from the raw profile data, that is, the variable values in $\mathbf{X}$.

An example of the former is shown in Table 5.1. Here, judgements about various brands of cola made by two subjects using a visual analogue scale with anchor points "same" (having a score of 0) and "different" (having a score of 100). In this example, the resulting rating for a pair of colas is a *dissimilarity*— low values indicate that the two colas are regarded as more alike than high values, and vice versa. A *similarity measure* would have been obtained had the anchor points been reversed, although similarities are often scaled to lie in the interval [0, 1]. An example of a proximity matrix arising from the basic data matrix is shown in Table 5.2. Here, the Euclidean distances between a number of pairs of countries have been calculated from the birth and death rates of each country.

The second technique that will be described in this chapter is *correspondence analysis*, which is essentially an approach to displaying the associations among a set of categorical variables in a type of *scatterplot* or *map*, thus allowing a visual examination of any structure or pattern in the data. Table 5.3, for example, shows a cross classification of 538 cancer patients by histological type, and by their response

Table 5.1 Dissimilarity Data for All Pairs of 10 Colas for Two Subjects

Subject 1:

					Cola number					
	1	2	3	4	5	6	7	8	9	10
1	0									
2	16	0								
3	81	47	0							
4	56	32	71	0						
5	87	68	44	71	0					
6	60	35	21	98	34	0				
7	84	94	98	57	99	99	0			
8	50	87	79	73	19	92	45	0		
9	99	25	53	98	52	17	99	84	0	
10	16	92	90	83	79	44	24	18	98	0

Subject 2:

					Cola number					
	1	2	3	4	5	6	7	8	9	10
1	0									
2	20	0								
3	75	35	0							
4	60	31	80	0						
5	80	70	37	70	0					
6	55	40	20	89	30	0				
7	80	90	90	55	87	88	0			
8	45	80	77	75	25	86	40	0		
9	87	35	50	88	60	10	98	83	0	
10	12	90	96	89	75	40	27	14	90	0

Table 5.2 Euclidean Distance Matrix Based on Birth and Death Rates for Five Countries

(1) Raw data

Country	Birth rate	Death rate
Algeria	36.4	14.6
France	18.2	11.7
Hungary	13.1	9.9
Poland	19.0	7.5
New Zealand	25.5	8.8

(2) Euclidean distance matrix

	Algeria	France	Hungary	Poland	New Zealand
Algeria	0.00				
France	18.43	0.00			
Hungary	23.76	5.41	0.00		
Poland	18.79	4.28	6.37	0.00	
New Zealand	12.34	7.85	12.45	6.63	0.00

Table 5.3 Hodgkin's Disease

Histological type	Response			
	Positive	Partial	None	Total
LP	74	18	12	104
NS	68	16	12	96
MC	154	54	58	266
LD	18	10	44	72
Total	314	98	126	538

to treatment three months after it had begun. A correspondence analysis of these data will be described later.

5.2 Multidimensional Scaling (MDS)

There are many methods of multidimensional scaling, and most of them are described in detail in Everitt and Rabe-Hesketh (1997). Here we shall concentrate on just one method, *classical multidimensional scaling*. Firstly, like all MDS techniques, classical scaling seeks to represent a proximity matrix by a simple geometrical model or map. Such a model is characterized by a set of points $x_1, x_2, \ldots, x_n$, in q dimensions, each point representing one of the stimuli of interest, and a measure of the distance between pairs of points. The objective of MDS is to determine both the dimensionality, q, of the model, and the n, q-dimensional coordinates, $x_1, x_2, \ldots, x_n$ so that the model gives a "good" fit for the observed proximities. Fit will often be judged by some numerical index that measures how well the proximities and the distances in the geometrical model match. In essence this simply means that the larger an observed dissimilarity between two stimuli (or the smaller their similarity), the further apart should be the points representing them in the final geometrical model.

The question now arises as to how we estimate q, and the coordinate values $x_1, x_2, \ldots, x_n$, from the observed proximity matrix? Classical scaling provides an answer to this question based on the work of Young and Householder (1938). To begin we must note that there is no unique set of coordinate values that give rise to these distances, since they are unchanged by shifting the whole configuration of points from one place to another, or by rotation or reflection of the configuration. In other words, we cannot uniquely determine either the location or the orientation of the configuration. The location problem is usually overcome by placing the mean vector of the configuration at the origin. The orientation problem means that any configuration derived can be subjected to an arbitrary orthogonal transformation. Such transformations can often be used to facilitate the interpretation of solutions as will be seen later.

The essential mathematical details of classical multidimensional scaling are given in Display 5.1.

Display 5.1
Mathematical Details of Classical Multidimensional Scaling

- To begin our account of the method we shall assume that the proximity matrix we are dealing with is a matrix of Euclidean distances derived from a raw data matrix, $\mathbf{X}$.
- In Chapter 1, we saw how to calculate Euclidean distances from $\mathbf{X}$. Multidimensional scaling is essentially concerned with the reverse problem: Given the distances (arrayed in the $n \times n$ matrix, $\mathbf{D}$) how do we find $\mathbf{X}$?
- To begin, define an $n \times n$ matrix $\mathbf{B}$ as follows

$$\mathbf{B} = \mathbf{XX}' \qquad \text{(a)}$$

- The elements of $\mathbf{B}$ are given by

$$b_{ij} = \sum_{k=1}^{q} x_{ik} x_{jk}. \qquad \text{(b)}$$

- It is easy to see that the squared Euclidean distances between the rows of $\mathbf{X}$ can be written in terms of the elements of $\mathbf{B}$ as

$$d_{ij}^2 = b_{ii} + b_{jj} - 2b_{ij}. \qquad \text{(c)}$$

- If the b's could be found in terms of the d's in the equation above, then the required coordinate value could be derived by factoring $\mathbf{B}$ as in (a).
- No unique solution exists unless a location constraint is introduced. Usually the center of the points $\bar{\mathbf{x}}$ is set at the origin, so that $\sum_{i=1}^{n} x_{ik} = 0$ for all k.
- These constraints and the relationship given in (b) imply that the sum of the terms in any row of $\mathbf{B}$ must be zero.
- Consequently, summing the relationship given in (c) over i, over j, and finally over both i and j, leads to the following series of equations:

$$\sum_{i=1}^{n} d_{ij}^2 = T + nb_{jj},$$

$$\sum_{i=1}^{n} d_{ij}^2 = nb_{ii} + T,$$

$$\sum_{i=1}^{n} \sum_{j=1}^{n} d_{ij}^2 = 2nT,$$

where $T = \sum_{i=1}^{n} b_{ii}$ is the trace of the matrix $\mathbf{B}$.
- The elements of $\mathbf{B}$ can now be found in terms of squared Euclidean distances as

$$b_{ij} = -\frac{1}{2} \left[d_{ij}^2 - d_{i.}^2 - d_{.j}^2 + d_{..}^2 \right],$$

where

$$d_{i.}^2 = \frac{1}{n} \sum_{j=1}^{n} d_{ij}^2,$$

$$d_{.j}^2 = \frac{1}{n} \sum_{i=1}^{n} d_{ij}^2,$$

$$d_{..}^2 = \frac{1}{n^2} \sum_{i=1}^{n} \sum_{j=1}^{n} d_{ij}^2.$$

- Having now derived the elements of **B** in terms of Euclidean distances, it remains to factor it to give the coordinate values.
- In terms of its singular value decomposition **B** can be written as

$$\mathbf{B} = \mathbf{V}\mathbf{\Lambda}\mathbf{V}',$$

where $\mathbf{\Lambda} = \mathrm{diag}[\lambda_1, \ldots, \lambda_n]$ is the diagonal matrix of eigenvalues of **B** and $\mathbf{V} = [\mathbf{V}_1, \ldots, \mathbf{V}_n]$, the corresponding matrix of eigenvectors, normalized so that the sum of squares of their elements is unity, that is, $\mathbf{V}_i'\mathbf{V}_i = 1$. The eigenvalues are assumed labeled such that $\lambda_1 \geq \lambda_2 \geq \cdots \geq \lambda_n$.
- When **D** arises from an $n \times q$ matrix of full rank, then the rank of **B** is q, so that the last $n - q$ of its eigenvalues will be zero.
- So **B** can be written as

$$\mathbf{B} = \mathbf{V}_1\mathbf{\Lambda}_1\mathbf{V}_1',$$

where $\mathbf{V}_1$ contains the first q eigenvectors and $\mathbf{\Lambda}_1$ the q nonzero eigenvalues.
- The required coordinate values are thus

$$\mathbf{X} = \mathbf{V}_1\mathbf{\Lambda}_1^{1/2}$$

where $\mathbf{\Lambda}_1^{1/2} = \mathrm{diag}[\lambda_1^{1/2}, \ldots, \lambda_p^{1/2}]$.
- The best fitting k-dimensional representation is given by the k eigenvectors of **B** corresponding to the k largest eigenvalues.
- The adequacy of the k-dimensional representation can be judged by the size of the criterion

$$P_k = \frac{\sum_{i=1}^{k} \lambda_i}{\sum_{i=1}^{n-1} \lambda_i}.$$

- Values of P_k of the order of 0.8 suggest a reasonable fit.
- When the observed dissimilarity matrix is not Euclidean, the matrix **B** is not positive-definite.
- In such cases some of the eigenvalues of **B** will be negative; correspondingly, some coordinate values will be complex numbers.

- If, however, **B** has only a small number of small negative eigenvalues, a useful representation of the proximity matrix may still be possible using the eigenvectors associated with the k largest positive eigenvalues.
- The adequacy of the resulting solution might be assessed using one of the following two criteria suggested by Mardia et al. (1979)

$$P_k^{(1)} = \frac{\sum_{i=1}^{k} |\lambda_i|}{\sum_{i=1}^{n} |\lambda_i|}$$

$$P_k^{(2)} = \frac{\sum_{i=1}^{k} \lambda_i^2}{\sum_{i=1}^{n} \lambda_i^2}$$

- Alternatively, Sibson (1979) recommends the following:
    1. *Trace criterion*: Choose the number of coordinates so that the sum of their positive eigenvalues is approximately equal to the sum of all the eigenvalues.
    2. *Magnitude criterion*: Accept as genuinely positive only those eigenvalues whose magnitude substantially exceeds that of the largest negative eigenvalue.

5.2.1 *Examples of Classical Multidimensional Scaling*

For our first example we will use the small set of multivariate data shown in Table 5.4, and the associated matrix of Euclidean distances will be our proximity matrix. To apply classical scaling to this matrix in R and S-PLUS® we can use the dist function to calculate the Euclidean distances combined with the cmdscale function to do the scaling

```
cmdscale(dist(x),k=5)
```

Here the five-dimensional solution (see Table 5.5) achieves complete recovery of the observed distance matrix. We can see this by comparing the original distances with those calculated from the scaling solution coordinates using the following R and S-PLUS code:

```
dist(x)- dist(cmdscale(dist(x), k=5)
```

The result is essentially a matrix of zeros.

The best fit in lower numbers of dimensions uses the coordinate values from the scaling solution in order from one to five. In fact, when the proximity matrix contains Euclidean distances derived from the raw data matrix, **X**, classical scaling can be shown to be equivalent to principal component analysis (see Chapter 3), with the derived coordinate values corresponding to the scores on the principal components derived from the covariance matrix. One result of this duality is the classical MDS

Table 5.4 Multivariate Data and Associated Euclidean Distances

(1) Data

$$X = \begin{pmatrix} 3 & 4 & 4 & 6 & 1 \\ 5 & 1 & 1 & 7 & 3 \\ 6 & 2 & 0 & 2 & 6 \\ 1 & 1 & 1 & 0 & 3 \\ 4 & 7 & 3 & 6 & 2 \\ 2 & 2 & 5 & 1 & 0 \\ 0 & 4 & 1 & 1 & 1 \\ 0 & 6 & 4 & 3 & 5 \\ 7 & 6 & 5 & 1 & 4 \\ 2 & 1 & 4 & 3 & 1 \end{pmatrix}$$

(2) Euclidean distances

$$D = \begin{pmatrix} 0.00 \\ 5.20 & 0.00 \\ 8.37 & 6.08 & 0.00 \\ 7.87 & 8.06 & 6.32 & 0.00 \\ 3.46 & 6.56 & 8.37 & 9.27 & 0.00 \\ 5.66 & 8.42 & 8.83 & 5.29 & 7.87 & 0.00 \\ 6.56 & 8.60 & 8.19 & 3.87 & 7.42 & 5.00 & 0.00 \\ 6.16 & 8.89 & 8.37 & 6.93 & 6.00 & 7.07 & 5.70 & 0.00 \\ 7.42 & 9.05 & 6.86 & 8.89 & 6.56 & 7.55 & 8.83 & 7.42 & 0.00 \\ 4.36 & 6.16 & 7.68 & 4.80 & 7.14 & 2.64 & 5.10 & 6.71 & 8.00 & 0.00 \end{pmatrix}$$

is often referred to as *principal coordinates analysis* (see Gower, 1966). The low-dimensional representation achieved by classical MDS for Euclidean distances (and that produced by principal component analysis) is such that the function ϕ given by

$$\phi = \sum_{r,s}^{n} (d_{rs}^2 - \hat{d}_{rs}^2)$$

is minimized. In this expression, d_{rs} is the Euclidean distance between observations r and s in the original q-dimensional space, and $\hat{d}_{rs}$ is the corresponding distance in

Table 5.5 Five-Dimensional Solution from Classical MDS Applied to the Distance Matrix in Table 5.4

	1	2	3	4	5
1	1.60	2.38	2.23	−0.37	0.12
2	2.82	−2.31	3.95	0.34	0.33
3	1.69	−5.14	−1.29	0.65	−0.05
4	−3.95	−2.43	−0.38	0.69	0.03
5	3.60	2.78	0.26	1.08	−1.26
6	−2.95	1.35	0.19	−2.82	0.12
7	−3.47	0.76	−0.30	1.64	−1.94
8	−0.35	2.31	−2.22	2.92	2.00
9	2.94	−0.01	−4.31	−2.51	−0.19
10	−1.93	0.33	1.87	−1.62	0.90

k-dimensional space ($k < q$) chosen for the classical scaling solution (equivalently the first k components).

Now let us look at an example involving distances that are not Euclidean and for this we shall use the data shown in Table 5.6 giving the airline distances between 10 U.S. cities and available as the dataframe `airline.dist`. These distances are not Euclidean since they relate essentially to journeys along the surface of a sphere. To apply classical scaling to these distances and to see the eigenvalues we can use the following R and S-PLUS code:

```
airline.mds<-cmdscale(airline.dist, k=9, eig=T)
airline.mds$eig
```

The eigenvalues are shown in Table 5.7. Some are negative for these non-Euclidean distances (and there are some small differences between R and S-PLUS after the fourth eigenvalue). We will assess how many coordinates we need to adequately represent the observed distance matrix using the criterion, $P_k^{(1)}$ in Display 5.1. The values of the criterion calculated from the eigenvalues in Table 5.7 for the one-dimensional and two-dimensional solutions are

$$P_1^{(1)} = 0.74, \qquad P_1^{(2)} = 0.93,$$
$$P_2^{(1)} = 0.91, \qquad P_2^{(1)} = 0.99.$$

These values suggest that the first two coordinates will give an adequate representation of the observed distances.

The plot of the two-dimensional coordinate values is obtained using

```
#
par(pty="s")
#use same limits for x and y axes
#
plot(airline.mds$points[,1],airline.mds$points[,2],
type="n",xlab="Coordinate 1",ylab="Coordinate 2",
xlim=c(-2000,1500), ylim=c(-2000,1500))
```

Table 5.6 Airline Distances Between 10 U.S. Cities

	Atla	Chic	Denv	Hous	LA	Mia	NY	SF	Seat	Wash
Atlanta	—	587	1212	701	1936	604	748	2139	218	543
Chicago	587	—	920	940	1745	1188	713	1858	1737	597
Denver	1212	920	—	879	831	1726	1631	949	1021	1494
Houston	701	940	879	—	1374	968	1420	1645	1891	1220
Los Angeles	1936	1745	831	1374	—	2338	2451	347	959	2300
Miami	604	1188	1726	968	2338	—	1092	2594	2734	923
New York	748	713	1631	1420	2451	1092	—	2571	2408	205
San Francisco	2139	1858	949	1645	347	2594	2571	—	678	2442
Seattle	218	1737	1021	1891	959	2734	2408	678	—	2329
Wash. D.C	543	597	1494	1220	2300	923	205	2442	2329	—

In dataframe `airline.dist`

Table 5.7 Eigenvalues and Eigenvectors Arising from Classical
Multidimensional Scaling Applied to Distance in Table 5.6

Eigenvalues	City	1	2
9.21×10^6	Atlanta	434.76	−724.22
2.20×10^6	Chicago	412.61	−55.04
1.08×10^6	Denver	−468.20	180.66
3.32×10^3	Houston	175.58	515.22
3.86×10^2	Los Angeles	−1206.68	465.64
-3.26×10^{-1}	Miami	1161.69	477.98
-9.30×10	New York	1115.56	−199.79
-2.17×10^3	San Francisco	−1422.69	308.66
-9.09×10^3	Seattle	−1221.54	−887.20
-1.72×10^6	Wash. D.C	1018.90	−81.90

```
text(airline.mds$points[,1],airline.mds$points[,2],
labels=row.names(airline.dist))
```

and is shown in Figure 5.1. (The coordinates obtained from R may have different
signs in which case some small amendments to the above code will be needed to
get the same diagram as Figure 5.1.)

Our last example of the use of classical multidimensional scaling will involve
the data shown in Table 5.8. These data show four measurements on male Egyptian
skulls from five epochs. The measurements are

MB: Maximum breadth
BH: Basibregmatic height

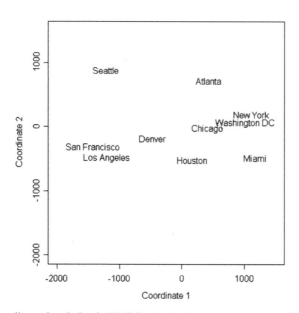

Figure 5.1 Two-dimensional classical MDS solution for airline distances from S-PLUS.

Table 5.8 Contents of Skull Dataframe. From *The Ancient Races of the Thebaid*, Arthur Thomson & R. Randall-Maciver (1905). By permission of Oxford University Press

	EPOCH	MB	BH	BL	NH
1	c4000BC	131	138	89	49
2	c4000BC	125	131	92	48
3	c4000BC	131	132	99	50
4	c4000BC	119	132	96	44
5	c4000BC	136	143	100	54
6	c4000BC	138	137	89	56
7	c4000BC	139	130	108	48
8	c4000BC	125	136	93	48
9	c4000BC	131	134	102	51
10	c4000BC	134	134	99	51
11	c4000BC	129	138	95	50
12	c4000BC	134	121	95	53
13	c4000BC	126	129	109	51
14	c4000BC	132	136	100	50
15	c4000BC	141	140	100	51
16	c4000BC	131	134	97	54
17	c4000BC	135	137	103	50
18	c4000BC	132	133	93	53
19	c4000BC	139	136	96	50
20	c4000BC	132	131	101	49
21	c4000BC	126	133	102	51
22	c4000BC	135	135	103	47
23	c4000BC	134	124	93	53
24	c4000BC	128	134	103	50
25	c4000BC	130	130	104	49
26	c4000BC	138	135	100	55
27	c4000BC	128	132	93	53
28	c4000BC	127	129	106	48
29	c4000BC	131	136	114	54
30	c4000BC	124	138	101	46
31	c3300BC	124	138	101	48
32	c3300BC	133	134	97	48
33	c3300BC	138	134	98	45
34	c3300BC	148	129	104	51
35	c3300BC	126	124	95	45
36	c3300BC	135	136	98	52
37	c3300BC	132	145	100	54
38	c3300BC	133	130	102	48
39	c3300BC	131	134	96	50
40	c3300BC	133	125	94	46
41	c3300BC	133	136	103	53
42	c3300BC	131	139	98	51
43	c3300BC	131	136	99	56
44	c3300BC	138	134	98	49
45	c3300BC	130	136	104	53
46	c3300BC	131	128	98	45
47	c3300BC	138	129	107	53
48	c3300BC	123	131	101	51
49	c3300BC	130	129	105	47
50	c3300BC	134	130	93	54
51	c3300BC	137	136	106	49
52	c3300BC	126	131	100	48
53	c3300BC	135	136	97	52

(Continued)

Table 5.8 (*Continued*)

	EPOCH	MB	BH	BL	NH
54	c3300BC	129	126	91	50
55	c3300BC	134	139	101	49
56	c3300BC	131	134	90	53
57	c3300BC	132	130	104	50
58	c3300BC	130	132	93	52
59	c3300BC	135	132	98	54
60	c3300BC	130	128	101	51
61	c1850BC	137	141	96	52
62	c1850BC	129	133	93	47
63	c1850BC	132	138	87	48
64	c1850BC	130	134	106	50
65	c1850BC	134	134	96	45
66	c1850BC	140	133	98	50
67	c1850BC	138	138	95	47
68	c1850BC	136	145	99	55
69	c1850BC	136	131	92	46
70	c1850BC	126	136	95	56
71	c1850BC	137	129	100	53
72	c1850BC	137	139	97	50
73	c1850BC	136	126	101	50
74	c1850BC	137	133	90	49
75	c1850BC	129	142	104	47
76	c1850BC	135	138	102	55
77	c1850BC	129	135	92	50
78	c1850BC	134	125	90	60
79	c1850BC	138	134	96	51
80	c1850BC	136	135	94	53
81	c1850BC	132	130	91	52
82	c1850BC	133	131	100	50
83	c1850BC	138	137	94	51
84	c1850BC	130	127	99	45
85	c1850BC	136	133	91	49
86	c1850BC	134	123	95	52
87	c1850BC	136	137	101	54
88	c1850BC	133	131	96	49
89	c1850BC	138	133	100	55
90	c1850BC	138	133	91	46
91	c200BC	137	134	107	54
92	c200BC	141	128	95	53
93	c200BC	141	130	87	49
94	c200BC	135	131	99	51
95	c200BC	133	120	91	46
96	c200BC	131	135	90	50
97	c200BC	140	137	94	60
98	c200BC	139	130	90	48
99	c200BC	140	134	90	51
100	c200BC	138	140	100	52
101	c200BC	132	133	90	53
102	c200BC	134	134	97	54
103	c200BC	135	135	99	50
104	c200BC	133	136	95	52
105	c200BC	136	130	99	55
106	c200BC	134	137	93	52
107	c200BC	131	141	99	55
108	c200BC	129	135	95	47
109	c200BC	136	128	93	54
110	c200BC	131	125	88	48

(*Continued*)

Table 5.8 (*Continued*)

	EPOCH	MB	BH	BL	NH
111	c200BC	139	130	94	53
112	c200BC	144	124	86	50
113	c200BC	141	131	97	53
114	c200BC	130	131	98	53
115	c200BC	133	128	92	51
116	c200BC	138	126	97	54
117	c200BC	131	142	95	53
118	c200BC	136	138	94	55
119	c200BC	132	136	92	52
120	c200BC	135	130	100	51
121	cAD150	137	123	91	50
122	cAD150	136	131	95	49
123	cAD150	128	126	91	57
124	cAD150	130	134	92	52
125	cAD150	138	127	86	47
126	cAD150	126	138	101	52
127	cAD150	136	138	97	58
128	cAD150	126	126	92	45
129	cAD150	132	132	99	55
130	cAD150	139	135	92	54
131	cAD150	143	120	95	51
132	cAD150	141	136	101	54
133	cAD150	135	135	95	56
134	cAD150	137	134	93	53
135	cAD150	142	135	96	52
136	cAD150	139	134	95	47
137	cAD150	138	125	99	51
138	cAD150	137	135	96	54
139	cAD150	133	125	92	50
140	cAD150	145	129	89	47
141	cAD150	138	136	92	46
142	cAD150	131	129	97	44
143	cAD150	143	126	88	54
144	cAD150	134	124	91	55
145	cAD150	132	127	97	52
146	cAD150	137	125	85	57
147	cAD150	129	128	81	52
148	cAD150	140	135	103	48
149	cAD150	147	129	87	48
150	cAD150	136	133	97	51

BL: Basialiveolar length
NH: Nasal height

We shall calculate Mahalanobis generalized distances (see Chapter 1) between each pair of epochs using the `mahalanobis` function, and apply classical scaling to the resulting distance matrix. In this calculation we shall use the following estimate of the assumed common covariance matrix $\mathbf{S}$,

$$\mathbf{S} = \frac{29\mathbf{S}_1 + 29\mathbf{S}_2 + 29\mathbf{S}_3 + 29\mathbf{S}_4 + 29\mathbf{S}_5}{149},$$

where $\mathbf{S}_1, \mathbf{S}_2, \ldots, \mathbf{S}_5$ are the covariance matrices of the data in each epoch. We shall then use the first two coordinate values to provide a map of the data showing the relationships between epochs. The necessary R and S-PLUS code is

```
labs<-rep(1:5,rep(30,5))
centers<-matrix(0,nrow=5,ncol=4)
S<-matrix(0,nrow=4,ncol=4)
#
for(i in 1:5) {
                centers[i,]<-apply(skulls[labs==i,-1],2,mean)
                S<-S+29*var(skulls[,-1])
}
#
S<-S/145
#
mahal<-matrix(0,5,5)
#
for(i in 1:5) {
    mahal[i,]<-mahalanobis(centers,centers[i,],S)
}
#
win.graph()
par(pty="s")
coords<-cmdscale(mahal)
#set up plotting area
xlim<-c(-1.5,1.5)
plot(coords,xlab="C1",ylab="C2",type="n",xlim=xlim,
    ylim=xlim,lwd=2)
text(coords,labels=c("c4000BC","c3300BC","c1850BC","c200BC",
    "cAD150"),lwd=3)
```

The resulting plot is shown in Figure 5.2.

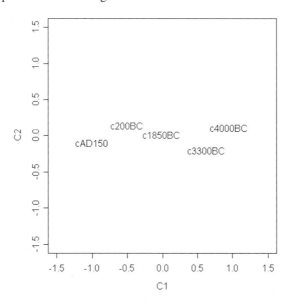

Figure 5.2 Two-dimensional solution from classical multidimensional scaling applied to the Mahalanobis distances between epochs for the skull data.

The scaling solution for the skulls data is essentially unidimensional, with this single-dimension time ordering the five epochs. There appears to be a change in the "shape" of the skulls over time with maximum breadth increasing and basialiveolar length decreasing.

5.3 Correspondence Analysis

Correspondence analysis has a relatively long history (see de Leeuw, 1985), but for a long period was only routinely used in France, largely due to the almost evangelical efforts of Benzécri (1992). But nowadays the method is used more widely and is often applied to supplement say a standard chi-squared test of independence for two categorical variables forming a contingency table.

Mathematically, correspondence analysis can be regarded as either:

- A method for decomposing the chi-squared statistic used to test for independence in a contingency table into components corresponding to different dimensions of the heterogeneity between its columns; or
- A method for simultaneously assigning a scale to rows and a separate scale to columns so as to maximize the correlation between the two scales.

Quintessentially however, correspondence analysis is a technique for displaying multivariate (most often bivariate) categorical data graphically, by deriving coordinates to represent the categories of both the row and column variables, which may then be plotted so as to display the pattern of association between the variables graphically.

In essence, correspondence analysis is nothing more than the application of classical multidimensional scaling to a specific type of distance suitable for categorical data, namely what is known as the *chi-squared distance*. Such distances are defined in Display 5.2. (A detailed account of correspondence analysis in which its similarity to principal components analysis is stressed is given in Greenacre, 1992.)

Display 5.2
Chi-Squared Distance

- The general contingency table in which there are r rows and c columns can be written as

		Columns				
		1	2		c	
	1	n_{11}	n_{12}		n_{1c}	$n_{1.}$
Rows	2	n_{21}				
	$\vdots$					
	r	n_{r1}			n_{rc}	$n_{r.}$
		$n_{.1}$			$n_{.c}$	N

using an obvious dot notation.

- From this we can construct tables of column proportions and row proportions given by

(a) Column proportions

	1		c
1	$p_{11} = n_{11}/n_{1.}$	$\cdots$	$p_{1c} = n_{1c}/n_{1.}$
2			
$\vdots$			
r	$p_{r1} = n_{r1}/n_{r.}$		$p_{rc} = n_{rc}/n_{r.}$

(b) Row proportions

	1		c
1	$p_{11} = n_{11}/n_{.1}$	$\cdots$	$p_{1c} = n_{1c}/n_{.1}$
2			
$\vdots$			
r	$p_{r1} = n_{r1}/n_{.1}$		$p_{rc} = n_{rc}/n_{.1}$

- The chi-squared distance between columns i and j is now defined as

$$d_{ij}^{(\text{cols})} = \sum_{k=1}^{r} \frac{1}{p_{k.}}(p_{ki} - p_{kj})^2$$

where

$$p_{k.} = \frac{n_{k.}}{N}$$

The chi-square distance is seen to be a weighted Euclidean distance based on column proportions. It will be zero if the two columns have the same values for these proportions. It can also be seen from the weighting factors $1/p_{k.}$ that rare categories of the column variable have a greater influence on the distance than common ones.

- A similar distance measure can be defined for rows i and j as

$$d_{ij}^{(\text{rows})} = \sum_{k=1}^{c} \frac{1}{p_{.k}}(p_{ik} - p_{jk})^2$$

where

$$p_{.k} = \frac{n_{.k}}{N}$$

- A correspondence analysis results from applying classical MDS to each distance matrix in turn and plotting say the first two coordinates for column categories and those for row categories on the same diagram, suitably labelled to differentiate the points representing row categories from those representing column categories.

- An explanation of how to interpret the derived coordinates is simplified by considering only a one-dimensional solution.
- When the coordinates for both row and columns category are large and positive (or both large and negative), it indicates a *positive* association between row i and column j; n_{ij} is greater than expected under the assumption of independence.
- Similarly, when the coordinates are both large in absolute values but have different signs, the corresponding row and column have a negative association; n_{ij} is less than expected under independence.
- Finally, when the product of the coordinates is near zero, the association between the row and column the column is low; n_{ij} is close to the value expected under independence.

As a simple introductory example, consider the data shown in Table 5.9 concerned with the influence of a girl's age on her relationship with her boyfriend. In this table each of 139 girls has been classified into one of three groups:

- No boyfriend;
- Boyfriend/no sexual intercourse;
- Boyfriend/sexual intercourse.

In addition, the age of each girl was recorded and used to divide the girls into five age groups. The calculation of the chi-squared distance measure can be illustrated using the proportions of girls in age groups 1 and 2 for each relationship type from Table 5.9:

$$\text{Chi-squared distance} = \sqrt{\frac{(0.68 - 0.64)^2}{0.55} + \frac{(0.26 - 0.27)^2}{0.24} + \frac{(0.06 - 0.09)^2}{0.21}}$$
$$= 0.09.$$

Table 5.9 The Influence of Age on Relationships with Boyfriends

	Age group				
	1(AG1)	2(AG2)	3(AG3)	4(AG4)	5(AG5)
No boyfriend (nbf)	21	21	14	13	8
(row percentage)	(68)	(64)	(58)	(42)	(40)
Boyfriend/no sexual intercourse (bfns)	8	9	6	8	2
(row percentage)	(26)	(27)	(25)	(26)	(10)
Boyfriend/sexual intercourse (bfs)	2	3	4	10	10
(row percentage)	(6)	(9)	(17)	(32)	(50)
Totals	31	33	24	31	20

NOTE: Age groups: (1) less than 16, (2) 16–17, (3) 17–18, (4) 18–19, (5) 19–20.

This is similar to ordinary Euclidean distance but differs in the division of each term by the corresponding average proportion. In this way the chi-squared distance measure compensates for the different levels of occurrence of the categories. (More formally, the choice of the chi-squared distance for measuring interprofile similarity can be justified as a way of standardizing variables under a multinomial or Poisson distributional assumption; see Greenacre, 1992.)

The complete set of chi-squared distances for all pairs of age groups can be arranged into the following matrix:

$$
dcols = \begin{array}{c} \\ \text{age group1} \\ \text{age group2} \\ \text{age group3} \\ \text{age group4} \\ \text{age group5} \end{array}
\begin{array}{ccccc}
1 & 2 & 3 & 4 & 5 \\
\left(\begin{array}{ccccc}
0.00 & 0.09 & 0.26 & 0.66 & 1.07 \\
0.09 & 0.00 & 0.19 & 0.59 & 1.01 \\
0.26 & 0.19 & 0.00 & 0.41 & 0.83 \\
0.66 & 0.59 & 0.41 & 0.00 & 0.51 \\
1.07 & 1.01 & 0.83 & 0.51 & 0.00
\end{array}\right)
\end{array}
$$

The corresponding matrix for rows is

$$
drows = \begin{array}{c} \\ \text{No boyfriend} \\ \text{Boyfriend/no sex} \\ \text{Boyfriend/sex} \end{array}
\begin{array}{ccc}
1 & 2 & 3 \\
\left(\begin{array}{ccc}
0.00 & 0.21 & 0.93 \\
0.21 & 0.00 & 0.93 \\
0.93 & 0.93 & 0.00
\end{array}\right)
\end{array}
$$

Applying classical MDS to each of these distance matrices gives the two-dimensional coordinates shown in Table 5.10. Plotting those with suitable labels and with the axes suitably scaled to reflect the greater variation on dimension one than on dimension two is achieved using the R and S-PLUS code:

```
r1<-cmdscale(dcols,eig=T)
c1<-cmdscale(drows,eig=T)
par(pty="s")
plot(r1$points,xlim=range(r1$points[,1],c1$points[,1]),
    ylim=range(r1$points[,1],c1$points[,1]),type="n",
xlab="Coordinate 1",ylab="Coordinate 2",lwd=2)
```

Table 5.10 Derived Correspondence Analysis Coordinates for Table 5.9

	x	y
No boyfriend	−0.304	−0.102
Boyfriend/no sexual intercourse	−0.312	0.101
Boyfriend/sexual intercourse	0.617	0.000
Age group 1	−0.402	0.062
Age group 2	−0.340	0.004
Age group 3	−0.153	−0.003
Age group 4	0.225	−0.152
Age group 5	0.671	0.089

```
text(r1$points,labels=c("AG1","AG2","AG3","AG4","AG5"),lwd=2)
text(c1$points,labels=c("nobf","bfns","bfs"),lwd=4)
abline(h=0,lty=2)
abline(v=0,lty=2)
```

to give Figure 5.3.

The points representing the age groups in Figure 5.4 give a two-dimensional representation of this distance, with the Euclidean distance between two points representing the chi-squared distance between the corresponding age groups. (This is similar for the points representing each type of relationship.) For a contingency table with I rows and J columns, it can be shown that the chi-squared distances can be represented *exactly* in $\min\{I - 1, J - 1\}$ dimensions; here since $I = 3$ and $J = 5$, this means that the Euclidean distances in Figure 5.4 will equal the corresponding chi-squared distances. For example, the correspondence analysis coordinates for age groups 1 and 2 taken from Table 5.10 are

Age group	x	y
1	−0.403	0.062
2	−0.339	0.004

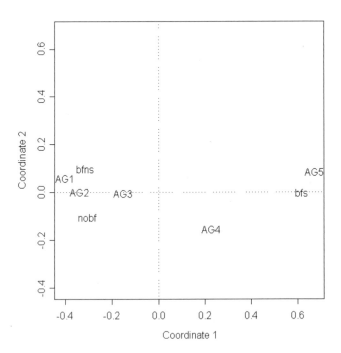

Figure 5.3 Classical multidimensional scaling of data in Table 5.9.

The corresponding Euclidean distance is calculated as

$$\sqrt{(-0.403 - 0.339)^2 + (0.062 - 0.004)^2} = 0.09$$

which agrees with the chi-squared distance between the two age groups calculated earlier.

When both I and J are greater than 3, an exact two-dimensional representation of the chi-squared distances is not possible. In such cases the derived two-dimensional coordinates will give only an approximate representation, and so the question of the adequacy of the fit will need to be addressed. In some of these cases more than two dimensions may be required to given an acceptable fit.

A correspondence analysis is interpreted by examining the positions of the row categories and the column categories as reflected by their respective coordinate values. The values of the coordinates reflect associations between the categories of the row variable and those of the column variable. If we assume that a two-dimensional solution provides an adequate fit, then row points that are close together indicate row categories that have similar profiles (conditional distributions) across the columns. Column points that are close together indicate columns with similar profiles (conditional distributions) down the rows. Finally, row points that are close to column points represent combinations that occur more frequently than would be expected under an independence model, that is, one in which the categories of the row variable are unrelated to the categories of the column variable.

Let's now look at two further examples of the application of correspondence analysis.

5.3.1 *Smoking and Motherhood*

Table 5.11 shows a set of frequency data first reported by Wermuth (1976). The data show the distribution of birth outcomes by age of mother, length of gestation, and whether or not the mother smoked during the prenatal period. We shall consider the data as a two-dimensional contingency table with four row categories and four column categories.

Table 5.11 Smoking and Motherhood

	Premature		Full term	
	Died in 1st year (pd)	Alive at year 1 (pa)	Died in 1st year (ftd)	Alive at year 1 (fta)
Young mothers				
Nonsmokers (YN)	50	315	24	4012
Smokers (YS)	9	40	6	459
Older mothers				
Nonsmokers (ONS)	41	147	14	1594
Smokers (YS)	4	11	1	124

The obvious question of interest for the data in Table 5.11 is whether or not a mother's smoking puts a newborn baby at risk. However, several other questions might also be of interest. Are smokers more likely to have premature babies? Are older mothers more likely to have premature babies? And how does smoking affect premature babies?

The chi-squared statistic for testing the independence of the two variables forming Table 5.11 takes the value 19.11 with 9 degrees of freedom; the associated p-value is 0.024. So it appears that "type" of mother is related to what happens to the newborn baby. We shall now examine how the results from a correspondence analysis can shed a little more light on this rather general finding. The relevant chi-squared distance matrices for these data are:

$$
dcols = \begin{array}{c c} & \begin{array}{c c c c} 1 & 2 & 3 & 4 \end{array} \\ \begin{array}{c} 1 \\ 2 \\ 3 \\ 4 \end{array} & \left(\begin{array}{c c c c} 0.00 & 0.30 & 0.27 & 0.37 \\ 0.30 & 0.00 & 0.23 & 0.07 \\ 0.27 & 0.23 & 0.00 & 0.28 \\ 0.37 & 0.07 & 0.28 & 0.00 \end{array} \right) \end{array}
$$

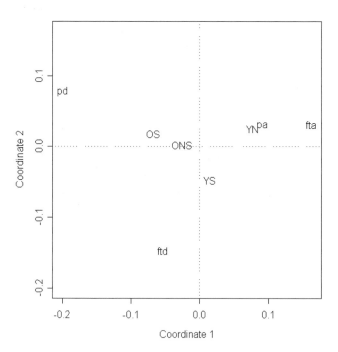

Figure 5.4 Two-dimensional solution for classical MDS applied to the motherhood and smoking data in Table 5.11.

$$drows = \begin{array}{c} \\ 1 \\ 2 \\ 3 \\ 4 \end{array} \begin{pmatrix} \begin{array}{cccc} 1 & 2 & 3 & 4 \\ 0.00 & 0.10 & 0.11 & 0.15 \\ 0.10 & 0.00 & 0.07 & 0.11 \\ 0.11 & 0.07 & 0.00 & 0.05 \\ 0.15 & 0.11 & 0.05 & 0.00 \end{array} \end{pmatrix}$$

Applying classical MDS and plotting the two-dimensional solution as above gives Figure 5.4. This diagram suggests that young mothers who smoke tend to have more full-term babies who then die in their first year, and older mothers who smoke have rather more than expected premature babies who die in the first year. It does appear that smoking is a risk factor for death in the first year of the baby's life, and that age is associated with length of gestation, with older mothers delivering more premature babies.

5.3.2 *Hodgkin's Disease*

The data shown in Table 5.3 were recorded during a study of Hodgkin's disease, a cancer of the lymph nodes; the study is described in Hancock et al. (1979). Each of 538 patients with the disease was classified by histological type, and by their response to treatment three months after it had begun. The histological classification is:

- lymphocyte predominance (LP),
- nodular sclerosis (NS),
- mixed cellularity (MC),
- lymphocyte depletion (LD).

The key question is, "What, if any, is the relationship between histological type and response to treatment?"

Here the chi-squared statistic takes the value 75.89 with 6 degrees of freedom. The associated p-value is very small. Clearly histological classification and response to treatment are related, but can correspondence analysis help in uncovering more about this association?

In this example the two-dimensional solution from applying classical MDS to the chi-squared distances gives a perfect fit. The resulting scatterplot is shown in Figure 5.5. The positions of the points representing histological classification and response to treatment in this diagram imply the following:

- Lymphocyte depletion tends to result in no response to treatment.
- Nodular sclerosis and lymphocyte predominance are associated with a positive response to treatment.
- Mixed cellularity tends to result in a partial response to treatment.

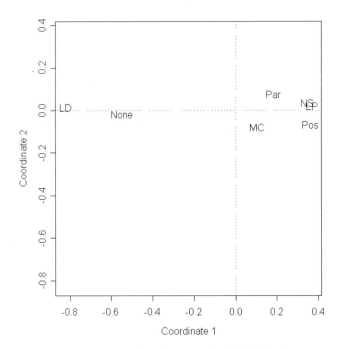

Figure 5.5 Classical MDS two-dimensional solution for Hodgkin's disease data.

5.4 Summary

Multidimensional scaling and correspondence analysis both aim to help in the understanding of particular types of data by displaying the data graphically. Multidimensional scaling applied to proximity matrices is often useful in uncovering the dimensions on which similarity judgments are made, and correspondence analysis often allows more insight into the pattern of relationships in a contingency table then a simple chi-squared test.

Exercises

5.1   What is mean by the *horseshoe effect* in multidimensional scaling solutions? (See Everitt and Rabe-Hesketh, 1997.) Create a similarity matrix as follows:

$$s_{ij} = 9 \text{ if } i = j,$$
$$= 8 \text{ if } 1 \leq |i - j| \leq 3,$$
$$\vdots$$
$$= 1 \text{ if } 2 \leq |i - j| \leq 2,$$
$$= 0 \text{ if } |i - j| > 25.$$

Table 5.12 Dissimilarity Matrix for a Set of Eight Legal Offenses

Offense	1	2	3	4	5	6	7	8
1	0							
2	21.1	0						
3	71.2	54.1	0					
4	36.4	36.4	36.4	0				
5	52.1	54.1	52.1	0.7	0			
6	89.9	75.2	36.4	54.1	53.0	0		
7	53.0	73.0	75.2	52.1	36.4	88.3	0	
8	90.1	93.2	71.2	63.4	52.1	36.4	73.0	0

Offenses: (1) assault and battery, (2) rape, (3) embezzlement, (4) perjury, (5) libel, (6) burglary, (7) prostitution, (8) receiving stolen goods.

Convert the resulting similarities into dissimilarities using $\delta_{ij} = \sqrt{s_{ii} + s_{jj} - 2s_{ij}}$ and find the two-dimensional configuration given by classical multidimensional scaling. The configuration should clearly show the horseshoe effect.

5.2   Show that classical multidimensional scaling applied to Euclidean distances calculated from a multivariate data matrix $\mathbf{X}$ is equivalent to principal components analysis, with the derived coordinate values corresponding to the scores on the principal components found from the covariance matrix of $\mathbf{X}$.

5.3   Write an S-PLUS (or R) function to calculate the chi-squared distance matrices for both rows and columns in a two-dimensional contingency table.

5.4   Table 5.12 summarizes data collected during a survey in which subjects were asked to compare a set of eight legal offenses, and to say for each one how unlike it was, in terms of seriousness, from the others. Each entry in the table shows the percentage of respondents who judged that the two offenses are very dissimilar. Find a two-dimensional scaling solution and try to interpret the dimensions underlying the subjects' judgements.

5.5   The data shown in Table 5.13 given the hair and eye color of a large number of people. Find the two-dimensional correspondence analysis solution for the data and plot the results.

Table 5.13 Hair Color and Eye Color of a Sample of Individuals

Eye color	Hair color				
	Fair	Red	Medium	Dark	Black
Light	688	116	584	188	4
Blue	326	38	241	110	3
Medium	343	84	909	412	26
Dark	98	48	403	681	81

Table 5.14 Suicides by Method, Sex, and Age

	Year							
	1970	1971	1972	1973	1974	1975	1976	1977
Shooting	15	15	31	17	42	49	38	27
Stabbing	95	113	94	125	124	126	148	127
Blunt instrument	23	16	34	34	35	33	41	41
Poison	9	4	8	3	5	3	1	4
Manual violence	47	60	54	70	69	66	70	60
Strangulation	43	45	43	53	51	63	47	51
Smothering/drowning	26	16	20	24	15	15	15	15

5.6 The data in Table 5.14 shows the methods by which victims of persons convicted for murder were killed between 1970 and 1977. How many dimensions would be needed for an *exact* correspondence analysis solution for these data? Use the first three correspondence analysis coordinates to plot a 3 × 3 scatterplot matrix (see Chapter 1). Interpret the results.

6
Cluster Analysis

6.1 Introduction

Cluster analysis is a generic term for a wide range of numerical methods for examining multivariate data with a view to uncovering or discovering groups or clusters of observations that are homogeneous and separated from other groups. In medicine, for example, discovering that a sample of patients with measurements on a variety of characteristics and symptoms actually consists of a small number of groups within which these characteristics are relatively similar, and between which they are different, might have important implications both in terms of future treatment and for investigating the aetiology of a condition. More recently cluster analysis techniques have been applied to microarray data (Alon et al., 1999) and image analysis (Everitt and Bullmore, 1999).

Clustering techniques essentially try to formalize what human observers do so well in two or three dimensions. Consider, for example, the scatterplot shown in Figure 6.1. The conclusion that there are two natural groups or clusters of dots is reached with no conscious effort or thought. Clusters are identified by the assessment of the relative distances between points and, in this example, the relative homogeneity of each cluster and the degree of their separation makes the task relatively simple.

Detailed accounts of clustering techniques are available in Everitt et al. (2001) and Gordon (1999). Here we concentrate on three types of clustering procedures.

- Agglomerative hierarchical methods;
- K-means type methods;
- Classification maximum likelihood methods.

6.2 Agglomerative Hierarchical Clustering

In a hierarchical classification the data are not partitioned into a particular number of classes or clusters at a single step. Instead the classification consists of a series of partitions that may run from a single "cluster" containing all individuals, to n clusters each containing a single individual. Agglomerative hierarchical clustering

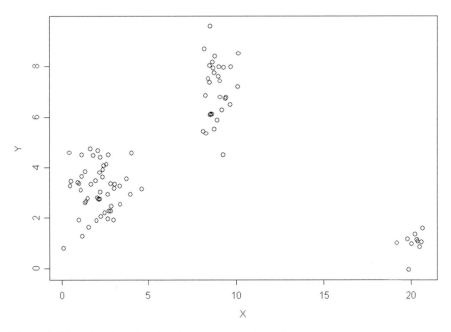

Figure 6.1 Bivariate data showing the presence of three clusters.

techniques produce partitions by a series of successive fusions of the n individuals into groups. Once made, however, such fusions are irreversible, so that when an agglomerative algorithm has placed two individuals in the same group, they cannot subsequently appear in different groups. Since all agglomerative hierarchical techniques ultimately reduce the data to a single cluster containing all the individuals, the investigator seeking the solution with the "best" fitting number of clusters will need to decide which division to choose. The problem of deciding on the "correct" number of clusters will be taken up later.

An agglomerative hierarchical clustering procedure produces a series of partitions of the data, $P_n, P_{n-1}, \ldots, P_1$. The first, P_n, consists of n single-member clusters, and the last, P_1, consists of a single group containing all n individuals. The basis operation of all methods is similar:

(START)   Clusters $C_1, C_2, \ldots, C_n$ each containing a single individual.
     (1)   Find the nearest pair of distinct clusters, say C_i and C_j, merge C_i and C_j, delete C_j and decrease the number of clusters by one.
     (2)   If number of clusters equals one then stop, else return to 1.

At each stage in the process the methods fuse individuals or groups of individuals which are closest (or most similar). The methods begin with an interindividual distance matrix (e.g., one containing Euclidean distances as defined in Chapter 1), but as groups are formed, distance between an individual and a group containing several individuals or between two groups of individuals will need to be calculated. How such distances are defined leads to a variety of different techniques; see the next subsection.

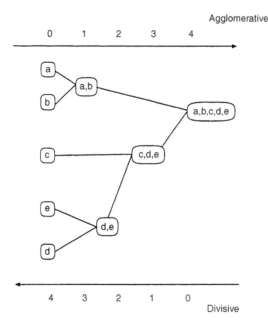

Figure 6.2 Example of a dendrogram. From *Finding Groups in Data: Introduction to Cluster Analysis*, Kaufman and Rousseeuw. Copyright © 1990. Reprinted with permission of John Wiley & Sons, Inc.

Hierarchic classifications may be represented by a two-dimensional diagram known as a *dendrogram*, which illustrates the fusions made a each stage of the analysis. An example of such a diagram is given in Figure 6.2. The structure of Figure 6.2 resembles an *evolutionary tree* (see Figure 6.3), and it is in biological applications that hierarchical classifications are most relevant and most justified

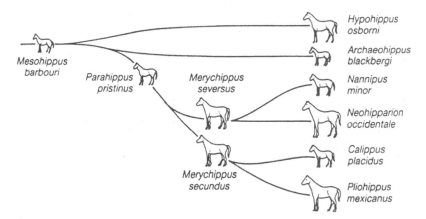

Figure 6.3 Evolutionary tree. From *Finding Groups in Data: Introduction to Cluster Analysis*, Kaufman and Rousseeuw. Copyright © 1990. Reprinted with permission of John Wiley & Sons, Inc.

(although this type of clustering has also been used in many other areas). According to Rohlf (1970), a biologist, "all things being equal," aims for a system of nested clusters. Hawkins et al. (1982), however, issue the following caveat: "users should be very wary of using hierarchic methods if they are not clearly necessary."

6.2.1 *Measuring Intercluster Dissimilarity*

Agglomerative hierarchical clustering techniques differ primarily in how they measure the distances between or similarity of two clusters (where a cluster may, at times, consist of only a single individual). Two simple intergroup measures are

$$d_{AB} = \min_{\substack{i \in A \\ i \in B}}(d_{ij}),$$

$$d_{AB} = \max_{\substack{i \in A \\ i \in B}}(d_{ij}),$$

where d_{AB} is the distance between two clusters A and B, and d_{ij} is the distance between individuals i and j. This could be Euclidean distance (see Chapter 1) or one of a variety of other distance measures; see Everitt et al., 2001, for details.

The first intergroup dissimilarity measure above is the basis of *single linkage* clustering, the second that of *complete linkage* clustering. Both these techniques have the desirable property that they are invariant under monotone transformations of the original interindividual dissimilarities or distances.

A further possibility for measuring intercluster distance or dissimilarity is

$$d_{AB} = \frac{1}{n_A n_B} \sum_{i \in A} \sum_{i \in B} d_{ij},$$

where n_A and n_B are the number of individuals in clusters A and B. This measure is the basis of a commonly used procedure known as *group average* clustering. All three intergroup measures described here are illustrated in Figure 6.4.

To illustrate the use of single linkage, complete linkage, and group average clustering we shall apply each method to the life expectancy data from the previous chapter (see Table 4.2). Here we assume that the eight life expectancies for each country are contained in the data frame life (see Chapter 4). The following R and S-PLUS code will calculate the Euclidean distance matrix for the countries, apply each of the clustering methods mentioned above, and then plot the resulting dendrograms, labelled with the country name:

R

```
#set up plotting area to take three side-by-side plots
country<-row.names(life)
par(mfrow=c(1,3))
#use dist to get Euclidean distance matrix, hclust to
#apply single linkage and plclust to plot dendrogram
plclust(hclust(dist(life),method="single"),
    labels=country,ylab="Distance")
```

```
title("(a) Single linkage")
plclust(hclust(dist(life),method="complete"),
    labels=country,ylab="Distance")
title("(b) Complete linkage")
plclust(hclust(dist(life),method="average"),
    labels=country,ylab="Distance")
title("(c) Average linkage")
```

S-PLUS

```
country<-row.names(life)
par(mfrow=c(1,3))
plclust(hclust(dist(life),method="connected"),
    labels=country,ylab="Distance")
title("(a) Single linkage")
plclust(hclust(dist(life),method="compact"),
    labels=country,ylab="Distance")
title("(b) Complete linkage")
plclust(hclust(dist(life),method="average"),
    labels=country,ylab="Distance")
title("(c) Average linkage")
```

The resulting diagram is shown in Figure 6.5.

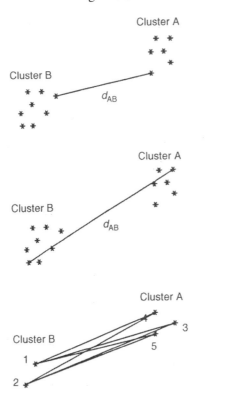

Figure 6.4 Intercluster distance measures.

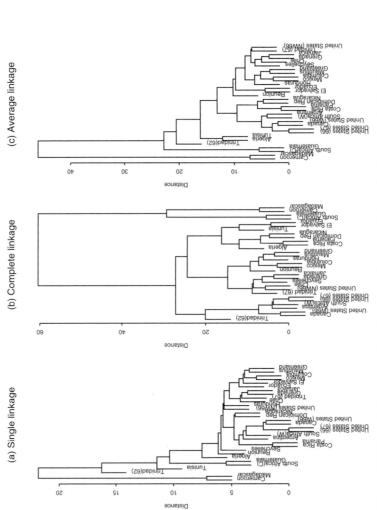

Figure 6.5 Single linkage, complete linkage, and average linkage dendrograms for the life expectancy data.

There are differences and similarities between the three dendrograms. Here we shall concentrate on the results given by complete linkage and we will examine the clustering found by "cutting" the complete linkage dendrogram at height 21 using the following R and S-PLUS® code:

R

```
four<-cutree(hclust(dist(life),method="complete"),h=21)
```

S-PLUS

```
four<-cutree(hclust(dist(life),method="compact"),h=21)
```

The resulting clusters in terms of country labels can be found from

```
#
country.clus<-lapply(1:5,function(nc)country[four==nc])
country.clus
```

The results from S-PLUS are shown in Table 6.1. (The group order differs in R, although the groups are the same.)

The means for the countries in each cluster can be found as follows:

```
country.mean<-lapply(1:5,function(nc)
   apply(life[four==nc,],2,mean))
country.mean
```

The results for the S-PLUS order of clusters are shown in Table 6.2. The S-PLUS clusters can be shown on a scatterplot matrix of the data using

```
pairs(life,panel=function(x,y) text(x,y,four))
```

The resulting plot is shown in Figure 6.6. This diagram suggests that the evidence for five distinct clusters in the data is not convincing.

Table 6.1 Clustering Solution from Complete Linkage

Cluster 1
South Africa (W), Canada, Trinidad (62), USA (66)
USA (W66), USA (67), Argentina
Cluster 2
Algeria, Tunisia, Costa Rica, Dominican Republic
El Salvador, Nicaragua, Panama, Ecuador
Cluster 3
Mauritius, Reunion, Seychelles, Greenland
Grenada, Honduras, Jamaica, Mexico
Trinidad (67), USA (NW66), Chile, Columbia
Cluster 4
Cameroon, Madagascar
Cluster 5
South Africa (C), Guatemala

Table 6.2 Mean Life Expectancies for the Five Clusters from Complete Linkage

	m0	m25	m50	m75	w0	w25	w50	w75
Cluster 1	66.4	48.0	22.9	7.9	72.7	50.7	27.7	9.7
Cluster 2	61.4	47.6	26.9	10.8	65.0	50.8	29.3	12.6
Cluster 3	60.1	42.8	22.0	7.6	64.9	46.8	25.3	9.7
Cluster 4	36.0	29.5	15.0	6.0	38.0	33.0	18.5	6.5
Cluster 5	49.5	39.5	21.0	8.0	53.0	42.0	23.0	8.0

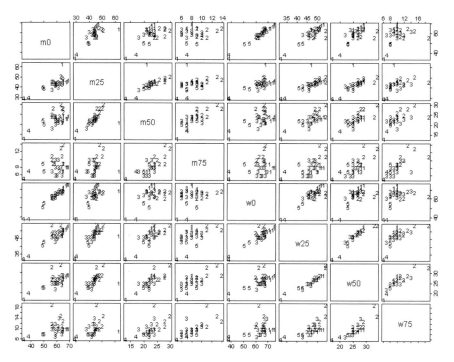

Figure 6.6 Scatterplot of life expectancy data showing five cluster solution from complete linkage.

6.3 *K*-Means Clustering

The k-means clustering technique seeks to partition a set of data into a specified number of groups, k, by minimizing some numerical criterion, low values of which are considered indicative of a "good" solution. The most commonly used approach, for example, is to try to find the partition of the n individuals into k groups, which minimizes the within-group sum of squares over all variables. The problem then appears relatively simple; namely, consider every possible partition of the n individuals into k groups, and select the one with the lowest within-group sum of squares.

Unfortunately, the problem in practice is not so straightforward. The numbers involved are so vast that complete enumeration of *every* possible partition remains impossible even with the fastest computer. To illustrate the scale of the problem:

n	k	Number of possible partitions
15	3	2, 375, 101
20	4	45, 232, 115, 901
25	8	690, 223, 721, 118, 368, 580
100	5	10^{68}

The impracticability of examining every possible partition has led to the development of algorithms designed to search for the minimum values of the clustering criterion by rearranging existing partitions and keeping the new one only if it provides an improvement. Such algorithms do not, of course, guarantee finding the global minimum of the criterion. The essential steps in these algorithms are as follows:

1. Find some initial partition of the individuals into the required number of groups. (Such an initial partition could be provided by a solution from one of the hierarchical clustering techniques described in the previous section.)
2. Calculate the change in the clustering criterion produced by "moving" each individual from its own to another cluster.
3. Make the change that leads to the greatest improvement in the value of the clustering criterion.
4. Repeat steps (2) and (3) until no move of an individual causes the clustering criterion to improve.

To illustrate the *k*-means approach with minimization of the within-clusters sum of squares criterion we shall apply it to the data shown in Table 6.3 which shows the chemical composition of 48 specimens of Romano-British pottery, determined by atomic absorption spectrophotometry, for nine oxides (Tubb et al., 1980).

Because the variables are on very different scales they will need to be standardized in some way before applying *k*-means clustering. In what follows we will divide each variable's values by the range of the variable. Assuming that the data are contained in a matrix `pottery.data`, this standardization can be applied in R and S-PLUS as follows:

```
rge<-apply(pottery.data,2,max)-apply(pottery.data,2,min)
pottery.dat<-sweep(pottery.data,2,rge,FUN="/")
```

The *k*-means approach can be used to partition the states into a prespecified number of clusters set by the investigator. In practice, solutions for a range of values for number of groups are found, but the question remains as to the "optimal" number of clusters for the data. A number of suggestions have been made as to how to tackle this question (see Everitt et al., 2001), but none is completely satisfactory. Here, we shall examine the value of the within-group sum of squares associated

Table 6.3 Results of Chemical Analyses of Romano British Pottery from Tubb et al. (1980) reprinted by kind permission of Blackwell Publishing

No	Kiln	Al_2O_3	Fe_2O_3	MgO	CaO	Na_2O	K_2O	TiO_2	MnO	BaO
1	1	18.8	9.52	2.00	0.79	0.40	3.20	1.01	0.077	0.015
2	1	16.9	7.33	1.65	0.84	0.40	3.05	0.99	0.067	0.018
3	1	18.2	7.64	1.82	0.77	0.40	3.07	0.98	0.087	0.014
4	1	16.9	7.29	1.56	0.76	0.40	3.05	1.00	0.063	0.019
5	1	17.8	7.24	1.83	0.92	0.43	3.12	0.93	0.061	0.019
6	1	18.8	7.45	2.06	0.87	0.25	3.26	0.98	0.072	0.017
7	1	16.5	7.05	1.81	1.73	0.33	3.20	0.95	0.066	0.019
8	1	18.0	7.42	2.06	1.00	0.28	3.37	0.96	0.072	0.017
9	1	15.8	7.15	1.62	0.71	0.38	3.25	0.93	0.062	0.017
10	1	14.6	6.87	1.67	0.76	0.33	3.06	0.91	0.055	0.012
11	1	13.7	5.83	1.50	0.66	0.13	2.25	0.75	0.034	0.012
12	1	14.6	6.76	1.63	1.48	0.20	3.02	0.87	0.055	0.016
13	1	14.8	7.07	1.62	1.44	0.24	3.03	0.86	0.080	0.016
14	1	17.1	7.79	1.99	0.83	0.46	3.13	0.93	0.090	0.020
15	1	16.8	7.86	1.86	0.84	0.46	2.93	0.94	0.94	0.20
16	1	15.8	7.65	1.94	0.81	0.83	3.33	0.96	0.112	0.019
17	1	18.6	7.85	2.33	0.87	0.39	3.17	0.98	0.081	0.018
18	1	16.9	7.87	1.83	1.31	0.53	3.09	0.95	0.092	0.023
19	1	18.9	7.58	2.05	0.83	0.13	3.29	0.98	0.072	0.015
20	1	18.0	7.50	1.94	0.69	0.12	3.14	0.93	0.035	0.017
21	1	17.8	7.28	1.92	0.81	0.18	3.15	0.90	0.067	0.017
22	2	14.4	7.00	4.30	0.15	0.51	4.25	0.79	0.160	0.019
23	2	13.8	7.08	3.43	0.12	0.17	4.14	0.77	0.144	0.020
24	2	14.6	7.09	3.88	0.13	0.20	4.36	0.81	0.124	0.019
25	2	11.5	6.37	5.64	0.16	0.14	3.89	0.69	0.087	0.009
26	2	13.8	7.06	5.34	0.20	0.20	4.31	0.71	0.101	0.021
27	2	10.9	6.26	3.47	0.17	0.22	3.40	0.66	0.109	0.010
28	2	10.1	4.26	4.26	0.20	0.18	3.32	0.59	0.149	0.017
29	2	11.6	5.78	5.91	0.18	0.16	3.70	0.65	0.082	0.015
30	2	11.1	5.49	4.52	0.29	0.30	4.03	0.63	0.080	0.016
31	2	13.4	6.92	7.23	0.28	0.20	4.54	0.69	0.163	0.017
32	2	12.4	6.13	5.69	0.22	0.54	4.65	0.70	0.159	0.015
33	2	13.1	6.64	5.51	0.31	0.24	4.89	0.72	0.094	0.017
34	3	11.6	5.39	3.77	0.29	0.06	4.51	0.56	0.110	0.015
35	3	11.8	5.44	3.94	0.30	0.04	4.64	0.59	0.085	0.013

(*Continued*)

Table 6.3 (*Continued*)

No	Kiln	Al$_2$O$_3$	Fe$_2$O$_3$	MgO	CaO	Na$_2$O	K$_2$O	TiO$_2$	MnO	BaO
36	4	18.3	1.28	0.67	0.03	0.03	1.96	0.65	0.001	0.014
37	4	15.8	2.39	0.63	0.01	0.04	1.94	1.29	0.001	0.014
38	4	18.0	1.50	0.67	0.01	0.06	2.11	0.92	0.001	0.016
39	4	18.0	1.88	0.68	0.01	0.04	2.00	1.11	0.006	0.022
41	4	20.8	1.51	0.72	0.07	0.10	2.37	1.26	0.002	0.016
42	5	17.7	1.12	0.56	0.06	0.06	2.06	0.79	0.001	0.013
43	5	18.3	1.14	0.67	0.06	0.05	2.11	0.89	0.006	0.019
44	5	16.7	0.92	0.53	0.01	0.05	1.76	0.91	0.004	0.013
45	5	14.8	2.74	0.67	0.03	0.05	2.15	1.34	0.003	0.015
46	5	19.1	1.64	0.60	0.10	0.03	1.75	1.04	0.007	0.018

with solutions for a range of values of k, the number of groups. As k increases this value will necessarily decrease but some "sharp" change may be indicative of the best solution. To obtain a plot of the within-group sum of squares for the one to six group solutions we can use the following R and S-PLUS code:

```
n<-length(pottery.dat[,1])
#find within group ss for all the data
wss1<-(n-1)*sum(apply(pottery.dat,2,var))
wss<-numeric(0)
#calculate within group ss for 2 to 6 group partitions given
   by k-means clustering
for(i in 2:6) {
        W<-sum(kmeans(pottery.dat,i)$withinss)
        wss<-c(wss,W)
}
wss<-c(wss1,wss)
plot(1:6,wss,type="l",xlab="Number of groups",
   ylab="Within groups sum of squares",lwd=2)
```

The resulting diagram is shown in Figure 6.7. The plot suggests looking at the two- or three-cluster solution. Details of the latter can be obtained using

```
pottery.kmean <- kmeans(pottery.dat, 3)
pottery.kmean
```

The output is shown in Table 6.4. The means from the code above are for the standardized data; to get the cluster means for the raw data we can use

```
lapply(1:3,function(nc)
   apply(pottery.dat[pottery.kmeans$cluster==nc,],2,mean))
```

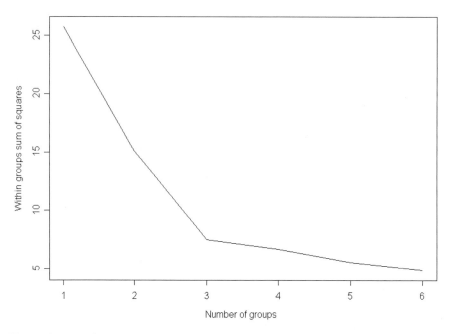

Figure 6.7 Plot of within-cluster sum of squares against number of clusters.

These means are also shown in Table 6.4. The means of each of the nine variables for each of the three clusters show that:

- Cluster three is characterized by a high aluminium oxide value and low iron oxide and calcium oxide values.
- Cluster two has a very high manganese oxide value and a high potassium oxide value.
- Cluster one has high calcium oxide value.

In addition to the chemical composition of the pots, the kiln site at which the pottery was found is known for these data (see Table 6.3). An archaeologist might be interested in assessing whether there is any association between the site and the distinct compositional groups found by the cluster analysis. To look at this we can cross-tabulate the kiln site against cluster label as follows:

```
table(kiln,pottery.kmean$cluster)
```

The resulting cross classification is shown in Table 6.5. Cluster 1 contains all 21 pots from kiln number one, cluster 2 contains pots from kilns 2 and 3, and cluster 3 pots from kilns 4 and 5. In fact, the five kiln sites are from three different regions defined by 1, (2, 3), (4, 5), so the clusters actually correspond to pots from three different regions.

Table 6.4 Details of Three-Group Solution for the Pottery Data

Means for standardized data
Centers:

	AL2O3	FE2O3	MGO	CAO	NA2O	K2O	TIO2	MNO	BAO
[1,]	1.5812	0.86379	0.274982	0.545958	0.43214	0.98817	1.20208	0.439153	1.2245
[2,]	1.1622	0.72184	0.713113	0.124585	0.28214	1.33371	0.87546	0.726190	1.1378
[3,]	1.6589	0.18744	0.095522	0.022674	0.06375	0.64363	1.30769	0.019753	1.1429

Clustering vector:
```
 [1] 1 1 1 1 1 1 1 1 1 1 1 1 1 1 1 1 1 1 1 1 1 2 2 2 2 2 2 2 2 2 2 2 2 2 2 3 3
[38] 3 3 3 3 3 3 3 3
```

Within cluster sum of squares:
```
[1] 3.1644 2.8748 1.4667
```

Cluster sizes:
```
[1] 21 14 10
```

Means for original data
[[1]]:

AL2O3	FE2O3	MGO	CAO	NA2O	K2O	TIO2	MNO	BAO
16.91905	7.428571	1.842381	0.9390476	0.3457143	3.102857	0.937619	0.07114286	0.01714286

[[2]]:

AL2O3	FE2O3	MGO	CAO	NA2O	K2O	TIO2	MNO	BAO
12.43571	6.207857	4.777857	0.2142857	0.2257143	4.187857	0.6828571	0.1176429	0.01592857

[[3]]:

AL2O3	FE2O3	MGO	CAO	NA2O	K2O	TIO2	MNO	BAO
17.75	1.612	0.64	0.039	0.051	2.021	1.02	0.0032	0.016

Table 6.5 Cross-Tabulation of
Cluster Label and Kiln

	1	2	3
1	21	0	0
2	0	12	0
3	0	2	0
4	0	0	5
5	0	0	5

6.4 Model-Based Clustering

The agglomerative hierarchical and k-means clustering methods described in the previous two sections are based largely in heuristic but intuitively reasonable procedures. But they are not based on formal models—those making problems such as deciding on a particular method, estimating the number of clusters, etc., particularly difficult. And, of course, without a reasonable model, formal inference is precluded. In practice, these may not be insurmountable objections to the use of the techniques since cluster analysis is essentially an "exploratory" tool. But model-based cluster methods do have some advantages, and a variety of possibilities have been proposed. The most successful approach has been that proposed by Scott and Symons (1971) and extended by Banfield and Raftery (1993) and Fraley and Raftery (2002), in which it is assumed that the population from the population from which the observations arise consists of c subpopulations, each corresponding to a cluster, and that the density of a q-dimensional observation from the jth subpopulation is $f_j(\mathbf{x}, \boldsymbol{\theta}_j)$ for some unknown vector of parameters, $\boldsymbol{\theta}_j$. They also introduce a vector $\boldsymbol{\gamma}' = [\gamma_1, \ldots, \gamma_n]$, where $\gamma_i = k$ if x_i is from the kth subpopulation; the γ_i label the subpopulation of each observation. The clustering problem now becomes that of choosing $\boldsymbol{\theta} = (\boldsymbol{\theta}_1, \boldsymbol{\theta}_2, \ldots, \boldsymbol{\theta}_c)$ and γ to maximize the likelihood function associated with such assumptions. This classification maximum likelihood procedure is described briefly in Display 6.1.

Display 6.1
Classification Maximum Likelihood

- Assume the population consists of c subpopulations, each corresponding to a cluster of observations, and that the density function of a q-dimensional observation from the jth subpopulation is $f_i(\mathbf{x}; \boldsymbol{\theta}_j)$ for some unknown vector of parameters, $\boldsymbol{\theta}_j$.
- Also, assume that $\boldsymbol{\gamma}' = [\gamma_1, \ldots, \gamma_n]$ gives the labels of the subpopulation to which each observation belongs. So $\gamma_i = j$ if $\mathbf{x}_i$ is from the jth population.

- The clustering problem becomes that of choosing $\boldsymbol{\theta}' = [\theta_1, \theta_2, \ldots, \theta_c]$ and $\boldsymbol{\gamma}$ to maximize the likelihood

$$L(\boldsymbol{\theta}, \boldsymbol{\gamma}) = \prod_{i=1}^{n} f_{\gamma_i}(\mathbf{x}_i; \boldsymbol{\theta}_{\gamma_i})$$

- If $f_j(\mathbf{x}; \boldsymbol{\theta}_j)$ is taken as a multivariate normal density with mean vector $\boldsymbol{\mu}_j$ and covariance matrix $\boldsymbol{\Sigma}_j$ this likelihood has the form

$$L(\boldsymbol{\theta}; \boldsymbol{\gamma}) = \text{const} \prod_{k=1}^{c} \prod_{i \in E_k} |\Sigma_k|^{1/2} \exp\left\{ -\frac{1}{2}(\mathbf{x}_i - \boldsymbol{\mu}_k)' \sum_k^{-1} (\mathbf{x}_i - \boldsymbol{\mu}_k) \right\}$$

where $E_j = \{i : \gamma_i = j\}$.

- The maximum likelihood estimator of $\boldsymbol{\mu}_j$ is $\bar{\mathbf{x}}_j = n_j^{-1} \Sigma_{i \in E_j} \mathbf{x}_i$ where n_j is the number of elements in E_j. Replacing $\boldsymbol{\mu}_j$ in (2) with this maximum likelihood estimator yields the following log-likelihood:

$$l(\theta, \gamma) = \text{const} - \frac{1}{2} \sum_{i=1}^{c} \text{trace}(\mathbf{W}_j \boldsymbol{\Sigma}_j^{-1} + n \log |\boldsymbol{\Sigma}_j|)$$

where $\mathbf{W}_j$ is the $p \times p$ matrix of sums of squares and cross-products of the variables for subpopulation j.

- Banfield and Raftery (1992) demonstrate the following:

  1. If $\boldsymbol{\Sigma}_k = \sigma^2 \mathbf{I}$ $(k = 1, 2, \ldots, c)$, then the likelihood is maximized by choosing γ to minimize trace $(\mathbf{W})$, where $\mathbf{W} = \Sigma_{k=1}^{c} \mathbf{W}_k$, that is, minimization of the written group sum of squares. Use of this criterion in a cluster analysis will tend to produce spherical clusters of largely equal sizes.
  2. If $\boldsymbol{\Sigma}_k = \boldsymbol{\Sigma}$ $(k = 1, 2, \ldots, c)$, then the likelihood is maximized by choosing γ to minimize $|\mathbf{W}|$, a clustering criterion discussed by Friedman and Rubin (1967) and Mariott (1982). Use of this criterion in a cluster analysis will tend to produce clusters with the same elliptical slope.
  3. If $\boldsymbol{\Sigma}_k$ is not constrained, the likelihood is maximized by choosing γ to minimize $\Sigma_{k=1}^{c} n_k \log |\mathbf{W}_k / n_k|$.

- Banfield and Raftery (1992) also consider criteria that allow the shape of clusters to be less constrained than with the minimization of trace $(\mathbf{W})$ and $|\mathbf{W}|$ criteria, but which remain more parsimonious then the completely unconstrained model. For example, constraining clusters to be spherical but not to have the same volume, or constraining clusters to have diagonal covariance matrices but allowing their shapes, sizes, and orientations to vary.

- The EM algorithm (see Dempster et al., 1977), is used for the maximum likelihood estimation; details are given in Fraley and Raftery (2002).
- Model selection is a combination of choosing the appropriate clustering model and the optimal number of clusters. A Bayesian approach is used (see Fraley and Raftery, 2002), using what is known as the *Bayesian Information Criterion* (BIC).

To illustrate this approach to clustering, we shall apply it to the data shown in Table 6.6. These data, taken with permission from Mayor and Frei (2003) give the values of three variables for the exoplanets discovered up to October 2002 (an exoplanet is a planet located outside the solar system). We assume the data are available as the data frame `planet.dat`.

R and S-PLUS functions for model-based clustering are available at http://www.stat.washington.edu/mclust. In R, the package can be installed from CRAN and then loaded in the usual way. Here we use the `Mclust` function since this selects both the most appropriate model for the data *and* the optimal number of groups based on the values of the BIC (see Display 6.1) computed over several models and a range of values for number of groups. The necessary code is

```
library(mclust)
planet.clus<-Mclust(planet.dat)
```

We can first examine a plot of BIC values using

```
plot(planet.clus,planet.dat)
```

and selecting the BIC option (option number 1). The resulting diagram is shown in Figure 6.8. In this diagram the numbers refer to different model assumptions about the shape of clusters:

1. Spherical, equal volume;
2. Spherical, unequal volume;
3. Diagonal equal volume, equal shape;
4. Diagonal varying volume, varying shape;
5. Ellipsoidal, equal volume, shape and orientation;
6. Ellipsoidal, varying volume, shape and orientation.

The BIC selects model 4 and three clusters as the best solution. This solution can be shown graphically on scatterplot matrix of the three variables constructed by using

Table 6.6 Data on Exoplanets, from Mayor et al. (2003), reprinted by kind permission of Cambridge University Press

Mass (in Jupiter mass)	Period (in Earth days)	Eccentricity
0.12	4.95	0
0.197	3.971	0
0.21	44.28	0.34
0.22	75.8	0.28
0.23	6.403	0.08
0.25	3.024	0.02
0.34	2.985	0.08
0.4	10.901	0.498
0.42	3.5097	0
0.47	4.229	0
0.48	3.487	0.05
0.48	22.09	0.3
0.54	3.097	0.01
0.56	30.12	0.27
0.68	4.617	0.02
0.685	3.524	0
0.76	2594	0.1
0.77	14.31	0.27
0.81	828.95	0.04
0.88	221.6	0.54
0.88	2518	0.6
0.89	64.62	0.13
0.9	1136	0.33
0.93	3.092	0
0.93	14.66	0.03
0.99	39.81	0.07
0.99	500.73	0.1
0.99	872.3	0.28
1	337.11	0.38
1	264.9	0.38
1.01	540.4	0.52
1.01	1942	0.4
1.02	10.72	0.044
1.05	119.6	0.35

(*Continued*)

Table 6.6 (*Continued*)

Mass (in Jupiter mass)	Period (in Earth days)	Eccentricity
1.12	500	0.23
1.13	154.8	0.31
1.15	2614	0
1.23	1326	0.14
1.24	391	0.4
1.24	435.6	0.45
1.282	7.1262	0.134
1.42	426	0.02
1.55	51.61	0.649
1.56	1444.5	0.2
1.58	260	0.24
1.63	444.6	0.41
1.64	406.0	0.53
1.65	401.1	0.36
1.68	796.7	0.68
1.76	903	0.2
1.83	454	0.2
1.89	61.02	0.1
1.9	6.276	0.15
1.99	743	0.62
2.05	241.3	0.24
0.05	1119	0.17
2.08	228.52	0.304
2.24	311.3	0.22
2.54	1089	0.06
2.54	627.34	0.06
2.55	2185	0.18
2.63	414	0.21
2.84	250.5	0.19
2.94	229.9	0.35
3.03	186.9	0.41
3.32	267.2	0.23
3.36	1098	0.22
3.37	133.71	0.511
3.44	1112	0.52
3.55	18.2	0.01

(*Continued*)

Table 6.6 (*Continued*)

Mass (in Jupiter mass)	Period (in Earth days)	Eccentricity
3.81	340	0.36
3.9	111.81	0.927
4	15.78	0.046
4	5360	0.16
4.12	1209.9	0.65
4.14	3.313	0.02
4.27	1764	0.353
4.29	1308.5	0.31
4.5	951	0.45
4.8	1237	0.515
5.18	576	0.71
5.7	383	0.07
6.08	1074	.011
6.292	71.487	0.1243
7.17	256	0.7
7.39	1582	0.478
7.42	116.7	0.4
7.5	2300	0.395
7.7	58.116	0.529
7.95	1620	0.22
8	1558	0.314
8.64	550.65	0.71
9.7	653.22	0.41
10	3030	0.56
10.37	2115.2	0.62
10.96	84.03	0.33
11.3	2189	0.34
11.98	1209	0.37
14.4	8.428	0.277
16.9	1739.5	0.228
17.5	256.03	0.429

```
plot(planet.clus,planet.dat)
```

and selecting the pairs option (option number 2). The plot is shown in Figure 6.9.
Mean vectors of the three clusters can be found from

```
planet.clus$mu
```

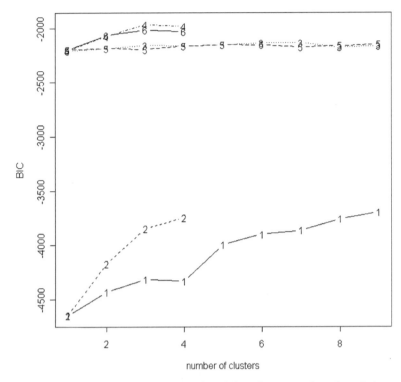

Figure 6.8 Plot of BIC values for a variety of models and a range of number of clusters.

and these are shown in Table 6.7. Cluster 1 consists of the "small" exoplanets (but still, on average, with a mass greater than Jupiter), with very short periods and eccentricities. The second cluster consists of large planets with very long periods and large eccentricities. The third cluster contains planets approximately the same mass as Jupiter, but with moderate periods and eccentricities.

6.5 Summary

Cluster analysis techniques provide a rich source of possible stategies for exploring complex multivariate data. They have been used widely in medical investigations; examples include Everitt et al. (1971) and Wastell and Gray (1987). Increasingly, model-based techniques such as finite mixture densities (see Everitt et al., 2001) and classification maximum likelihood, as described in this chapter, are superseding older methods, such as the single linkage, complete linkage, and average linkage methods described in Section 6.2. Two recent references are Fraley and Raftery (1998, 1999).

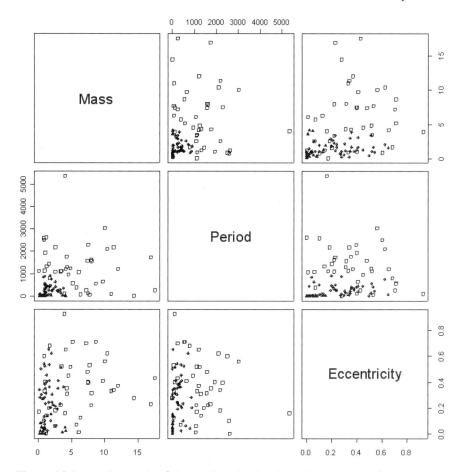

Figure 6.9 Scatterplot matrix of planets data showing three cluster solution from `Mclust`.

Table 6.7 Means for the Three-Group Solution for the Exoplanets Data

	Mass	Period	Eccentricity
Cluster 1: $n = 19$	1.16	6.45	0.035
Cluster 2: $n = 41$	5.81	1263.01	0.363
Cluster 3: $n = 15$	1.54	303.82	0.308

Exercises

6.1 Show that the intercluster distances used by single linkage, complete linkage, and group average clustering satisfy the following formula:

$$d_{k(ij)} = \alpha_i d_{ki} + \alpha_j d_{kj} + \gamma |d_{ki} - d_{kj}|,$$

where

$$\alpha_i = \alpha_j, \gamma = -\frac{1}{2} \qquad \text{(single linkage)},$$

$$\alpha_i = \alpha_j, \gamma = \frac{1}{2} \qquad \text{(complete linkage)},$$

$$a_i = \frac{n_i}{n_i + n_j}, \alpha_j = \frac{n_j}{n_i + n_j}, \gamma = 0 \qquad \text{(group average)}.$$

$(d_{k(ij)}$ is the distance between a group k and a group (ij) formed by the fusion of groups i and j, and d_{ij} is the distance between groups i and j; n_i and n_j are the number of observations in groups i and j.)

6.2 Ward (1963) proposed an agglomerative hierarchical clustering procedure in which, at each step, the union of every possible pair of clusters is considered and the two clusters whose fusion results in the minimum increase in an error sum-of-squares criterion, ESS, are combined. For a single variable, ESS for a group with n individuals is simply $ESS = \sum_{i=1}^{n}(x_i - \bar{x})^2$.

(a) If ten individuals with variable values {2, 6, 5, 6, 2, 2, 2, 0, 0, 0} are considered as a single group, calculate ESS. If the individuals are grouped into two groups with individuals 1, 5, 6, 7, 8, 9, 10 in one group and individuals 2, 3, 4 in the other, what does ESS become?

(b) Can you fit Ward's method into the general equation given in Exercise 5.1?

6.3 Reanalyze the pottery data using Mclust. To what model in Mclust does the k-mean approach approximate?

6.4 Construct a three-dimensional drop-line scatterplot of the planets data in which the points are labelled with a suitable cluster label.

6.5 Reanalyze the life expectancy data by clustering the countries on the basis on differences between the life expectancies of men and women at corresponding ages.

7
Grouped Multivariate Data: Multivariate Analysis of Variance and Discriminant Function Analysis

7.1 Introduction

Investigators in many disciplines frequently collect multivariate data on samples from different populations. In Chapter 5, for example, a set of data was introduced in which an archaeologist had made four measurements on Egyptian skulls from five different epochs. A variety of questions might be asked about such data and, correspondingly, there are a variety of (overlapping) approaches to their analysis. In many examples the prime interest will be in assessing whether the populations involved have different mean vectors on the measurements taken. For this, multivariate analogues of the familiar univariate t-test, *Hotelling's T^2*, or analysis of variance, *multivariate analysis of variance*, are available. A further question that is often of interest for grouped multivariate data is whether or not it is possible to use the measurements made to construct a classification rule derived from the original observations (the *training set*) that will allow new individuals having the same set of measurements, but no group label, to be allocated to a group in such a way that misclassifications are minimized. The relevant technique is now some form of *discriminant function analysis*.

In the next section we consider both the inference and classification questions for the two-group situation, and then in Section 7.3 move on to discuss data sets where there are more than two groups.

7.2 Two Groups: Hotellings T^2 Test and Fisher's Linear Discriminant Function Analysis

7.2.1 *Hotellings T^2 Test*

The data shown in Table 7.1 were originally collected by Colonel L.A. Waddell in southeastern and eastern Tibet. According to Morant (1923), the data consist of two groups of skulls: group one (type I), skulls 1–17, found in graves in Sikkim and neighboring areas of Tibet; group two (type II) consisting of the remaining 15 skulls picked up on battlefield in the Lhasa district and believed to be those of native

Table 7.1 Tibetan Skull Data (all measurements in mm). From Morant, G.M., *A First Study of the Tibetan Skull*, in Biometrika, Vol. 14, 1923, pp 193–260, by permission of the Biometrika Trustees

Obs	Length	Breadth	Height	Fheight	Fbreadth	Type
1	190.5	152.5	145.0	73.5	136.5	1
2	172.5	132.0	125.5	63.0	121.0	1
3	167.0	130.0	125.5	69.5	119.5	1
4	169.5	150.5	133.5	64.5	128.0	1
5	175.0	138.5	126.0	77.5	135.5	1
6	177.5	142.5	142.5	71.5	131.0	1
7	179.5	142.5	127.5	70.5	134.5	1
8	179.5	138.0	133.5	73.5	132.5	1
9	173.5	135.5	130.5	70.0	133.5	1
10	162.5	139.0	131.0	62.0	126.0	1
11	178.5	135.0	136.0	71.0	124.0	1
12	171.5	148.5	132.5	65.0	146.5	1
13	180.5	139.0	132.0	74.5	134.5	1
14	183.0	149.0	121.5	76.5	142.0	1
15	169.5	130.0	131.0	68.0	119.0	1
16	172.0	140.0	136.0	70.5	133.5	1
17	170.0	126.5	134.5	66.0	118.5	1
18	182.5	136.0	138.5	76.0	134.0	2
19	179.5	135.0	128.5	74.0	132.0	2
20	191.0	140.5	140.5	72.5	131.5	2
21	184.5	141.5	134.5	76.5	141.5	2
22	181.0	142.0	132.5	79.0	136.5	2
23	173.5	136.5	126.0	71.5	136.5	2
24	188.5	130.0	143.0	79.5	136.0	2
25	175.0	153.0	130.0	76.0	134.0	2
27	200.0	139.5	143.5	82.5	146.0	2
28	185.0	134.5	140.0	81.5	137.0	2
29	174.5	143.5	132.5	74.0	136.5	2
30	195.5	144.0	138.5	78.5	144.0	2
31	197.0	131.5	135.0	80.5	139.0	2
32	182.5	131.0	135.0	68.5	136.0	2

In dataframe `Tibet`.

soldiers from the eastern province of Khans. These skulls were of particular interest since it was thought at the time that Tibetans from Khans might be survivors of a particular fundamental human type, unrelated to the Mongolian and Indian types that surrounded them.

On each of the 32 skulls the following five measurements, all in millimeters, were recorded:

x_1: greatest length of skull (length),
x_2: greatest horizontal breadth of skull (breadth),
x_3: height of skull (height),
x_4: upper face height (fheight),
x_5: face breadth, between outermost points of cheek bones (fbreadth).

We assume the data are available as the dataframe `Tibet`.

The first task to carry out on these data is to test the hypothesis that the five-dimensional mean vectors of skull measurements are the same in the two populations from which the samples arise. For this we will use the multivariate analogue of Student's independent samples t-test, known as Hotelling's T^2 test, a test described in Display 7.1.

Display 7.1
Hotelling's T^2 Test

- If there are q variables, the null hypothesis is that the means of the variables in the first population equal the means of the variables in the second population.
- If μ_1 and μ_2 are the mean vectors of the two populations the null hypothesis can be written as

$$H_0: \mu_1 = \mu_2.$$

- The test statistic T^2 is defined as

$$T^2 = \frac{n_1 n_2}{n_1 + n_2} D^2,$$

where n_1 and n_2 are the sample sizes in each group and D^2 is the generalized distance introduced in Chapter 1, namely

$$D^2 = (\bar{\mathbf{x}}_1 - \bar{\mathbf{x}}_2)' \mathbf{S}^{-1} (\bar{\mathbf{x}}_1 - \bar{\mathbf{x}}_2),$$

where $\bar{\mathbf{x}}_1$ and $\bar{\mathbf{x}}_2$ are the two sample mean vectors and $\mathbf{S}$ is the estimate of the assumed common covariance matrix of the two populations, calculated from the two sample covariance matrix, $\mathbf{S}_1$ and $\mathbf{S}_2$ as

$$\mathbf{S} = \frac{(n_1 - 1)\mathbf{S}_1 + (n_2 - 1)\mathbf{S}_2}{n_1 + n_2 - 2}.$$

- Note that the form of the test statistic in the multivariate case is very similar to that for the univariate independent samples t-test, involving a difference between "means" (here mean vectors), and an assumed common "variance" (here a covariance matrix).

- Under H_0 (and when the assumptions given below hold), the statistic F given by

$$F = \frac{(n_1 + n_2 - q - 1)T^2}{(n_1 + n_2 - 2)q}$$

has a Fisher's F-distribution with q and $n_1 + n_2 - q - 1$ degrees of freedom.
- The T^2 test is based on the following assumptions:

  1. In each population the variables have a multivariate normal distribution.
  2. The two populations have the same covariance matrix.
  3. The observations are independent.

As an exercise we will apply Hotelling's T^2 test to the skull data using the following R and S-PLUS® code, although we could also use the manova function as we shall see later:

```
attach(Tibet)
m1<-apply(Tibet[Type==1,-6],2,mean)
m2<-apply(Tibet[Type==2,-6],2,mean)
l1<-length(Type[Type==1])
l2<-length(Type[Type==2])
x1<-Tibet[Type==1,-6]
x2<-Tibet[Type==2,-6]
S123<-((l1-1)*var(x1)+(l2-1)*var(x2))/(l1+l2-2)
T2<-t(m1-m2)%*%solve(S123)%*%(m1-m2)
Fstat<-(l1+l2-5-1)*T2/(l1+l2-2)*5
pvalue<-1-pf(Fstat,5,26)
```

Hotelling's T^2 takes the value 3.50 with the corresponding F statistic being 15.17 with 5 and 26 degrees of freedom. The associated p-value is very small, and we can conclude that there is strong evidence that the mean vectors of the two groups differ.

It might be thought that the results of Hotelling's T^2 test would simply reflect those that would be obtained using a series of univariate t-tests, in the sense that if no significant differences are found by the separate t-tests, then the T^2 test will inevitably lead to acceptance of the null hypothesis that the population mean vectors are equal. And, on the other hand, if any significant difference is found when using the t-tests on the individual variables, then the T^2 statistic must also lead to a significant result. But these speculations are not correct (if they were, the T^2 test would be a waste of time). It is entirely possible to find no significant difference for each separate t-test, but a significant result for the T^2 test, and vice versa. An illustration of how this can happen in the case of two variables is shown in Display 7.2.

Display 7.2

Univariate and Multivariate Tests for Equality of Means for Two Variables

- Suppose we have a sample of n observations on two variables, x_1 and x_2 and we wish to test whether the population means of the two variables μ_1 and μ_2 are both zero.
- Assume the mean and standard deviation of the x_1 observations are $\bar{x}_1$ and s_1, respectively, and of the x_2 observations, $\bar{x}_2$ and s_2.
- If we test separately whether each mean takes the value zero, then we would use two t tests. For example, to test $\mu_1 = 0$ against $\mu_1 \neq 0$ the appropriate test statistic is

$$t = \frac{\bar{x}_1 - 0}{s_1\sqrt{n}}.$$

- The hypothesis $\mu_1 = 0$ would be rejected at the α percent level of significance if $t < -t_{100\left(1-\frac{1}{2}\alpha\right)}$ or $t > t_{100\left(1-\frac{1}{2}\alpha\right)}$; that is, if $\bar{x}_1$ fell outside the interval $\left[-s_1 t_{100\left(1-\frac{1}{2}\alpha\right)}\middle/\sqrt{n}\right]$ where $t_{100\left(1-\frac{1}{2}\alpha\right)}$ is the $100\left(1-\frac{1}{2}\alpha\right)$ percent point of the t distribution with $n - 1$ degrees of freedom. Thus the hypothesis would *not* be rejected if $\bar{x}_1$ fell *within* this interval.
- Similarly, the hypothesis $\mu_2 = 0$ for the variable x_2 would not be rejected if the mean, $\bar{x}_2$, of the x_2 observations fell within a corresponding interval with s_2 substituted for s_1.
- The multivariate hypothesis $[\mu_1, \mu_2] = [0, 0]$ would therefore not be rejected if *both* these conditions were satisfied.
- If we were to plot the point $(\bar{x}_1, \bar{x}_2)$ against rectangular axes, the area within which the point could like and the multivariate hypothesis not rejected is given by the rectangle $ABCD$ of the diagram below, where AB and DC are of length $2s_1 t_{100\left(1-\frac{1}{2}\alpha\right)}\sqrt{n}$ while AD and BC are of length $2s_2 t_{100\left(1-\frac{1}{2}\alpha\right)}\sqrt{n}$.

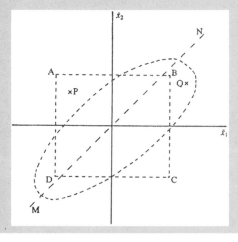

- Thus, a sample that gave the means $(\bar{x}_1, \bar{x}_2)$ represented by the point P would lead to acceptance of the multivariate hypothesis.
- Suppose, however, that the variables x_1 and x_2 are moderately highly correlated. Then all points (x_1, x_2) and hence $(\bar{x}_1, \bar{x}_2)$ should lie reasonably close to the straight line MN through the origin marked on the diagram.
- Hence samples consistent with the multivariate hypothesis should be represented by points $(\bar{x}_1, \bar{x}_2)$ that lie within a region encompassing the line MN. When we take account of the nature of the variation of bivariate normal samples that include correlation, this region can be shown to be an ellipse such as that marked on the diagram. The point P is *not* consistent with this region and, in fact, should be *rejected* for this sample.
- Thus, the inference drawn from the two separate univariate tests conflicts with the one drawn from a single multivariate test, and it is the wrong inference.
- A sample giving the $(\bar{x}_1, \bar{x}_2)$ values represented by point Q would give the other type of mistake, where the application of two separate univariate tests leads to the rejection of the null hypothesis, but the correct multivariate inference is that the hypothesis should *not* be rejected. (This explanation is taken with permission from Krzanowski, 1988.)

Having produced evidence that the mean vectors of skull types I and II are not the same, we can move on to the classification aspect of grouped multivariate data.

7.2.2 Fisher's Linear Discriminant Function

Suppose a further skull is uncovered whose origin is unknown, that is, we do not know if it is type I or type II. How might we use the original data to construct a classification rule that will allow the new skull to be classified as type I or II based on the same five measurements taken on the skulls in Table 7.1? The answer was provided by Fisher (1936) who approached the problem by seeking a linear function of the observed variables that provides maximal separation, in a particular sense, between the two groups. Details of Fisher's *linear discriminant function* are given in Display 7.3.

Display 7.3
Fisher's Linear Discriminant Function

- The aim is to find a way of classifying observations into one of two known groups using a set of variables, $x_1, x_2, \ldots, x_q$:
- Fisher's idea was to find a linear function z of the variables. $x_1, x_2, \ldots, x_q$;

$$z = a_1 x_1 + a_2 x_2 + \cdots + a_q x_q,$$

such that the ratio of the between-group variance of z to its within-group variance is maximized.

- The coefficients $\mathbf{a}' = [a_1, \ldots, a_q]$ have therefore to be chosen so that V, given by

$$V = \frac{\mathbf{a}'\mathbf{B}\mathbf{a}}{\mathbf{a}'\mathbf{S}\mathbf{a}},$$

is maximized, where $\mathbf{S}$ is the pooled within-group covariance matrix, and $\mathbf{B}$ is the covariance matrix of group means; explicitly,

$$\mathbf{S} = \frac{1}{n-2} \sum_{i=1}^{2} \sum_{j=1}^{n_i} (\mathbf{x}_{ij} - \bar{\mathbf{x}}_j)(\mathbf{x}_{ij} - \bar{\mathbf{x}}_j)',$$

$$\mathbf{B} = \sum_{i=1}^{2} n_i (\bar{\mathbf{x}}_i - \bar{\mathbf{x}})(\bar{\mathbf{x}}_i - \bar{\mathbf{x}})',$$

where $\mathbf{x}'_{ij} = [x_{ij1}, x_{ij2}, \ldots, x_{ijq}]$ represents the set of q variable values for the jth individual in group i, $\bar{\mathbf{x}}_j$ is the mean vector of the jth group, and $\bar{\mathbf{x}}$ is the mean vector of all observations. The number of observations in each group is n_1 and n_2, with $n = n_1 + n_2$.

- The vector $\mathbf{a}$ that maximizes V is given by the solution of

$$(\mathbf{B} - \lambda\mathbf{S})\mathbf{a} = 0.$$

- In the two-group situation, the *single* solution can be shown to be

$$\mathbf{a} = \mathbf{S}^{-1}(\bar{\mathbf{x}}_1 - \bar{\mathbf{x}}_2).$$

- The allocation rule is now to allocate an individual with discriminate score z to group 1 if $z > (\bar{z}_1 + \bar{z}_2)/2$, where $\bar{z}_1$ and $\bar{z}_2$ are the mean discriminant scores in each group. (We are assuming that the groups are labelled such that $\bar{z}_1 > \bar{z}_2$.)

- Fisher's discriminant function also arises from assuming that the observations in group one have a multivariate normal distribution with mean vector $\boldsymbol{\mu}_1$ and covariance matrix $\boldsymbol{\Sigma}$ and those in group two have a multivariate distribution with mean vector $\boldsymbol{\mu}_2$ and, again, covariance matrix $\boldsymbol{\Sigma}$, and assuming that an individual with vector of scores $\mathbf{x}$ is allocated to group one if

$$MVN(\mathbf{x}, \boldsymbol{\mu}_1, \boldsymbol{\Sigma}) > MVN(\mathbf{x}, \boldsymbol{\mu}_2, \boldsymbol{\Sigma}),$$

where MVN is shorthand for the multivariate normal density function.

- Substituting sample values for population rules leads to the same allocation rule as that given above.

- The above is only valid if the prior probabilities of being in each group are assumed to be the same.

- When the prior probabilities are not equal the classification rule changes; for details, see Everitt and Dunn (2001).

The description given in Display 7.3 can be translated into R and S-PLUS code as follows:

```
m1<-apply(Tibet[Type==1,-6],2,mean)
m2<-apply(Tibet[Type==2,-6],2,mean)
l1<-length(Type[Type==1])
l2<-length(Type[Type==2])
x1<-Tibet[Type==1,-6]
x2<-Tibet[Type==2,-6]
S123<-((l1-1)*var(x1)+(l2-1)*var(x2))/(l1+l2-2)
a<-solve(S123)%*%(m1-m2)
z12<-(m1%*%a+m2%*%a)/2
```

This leads to the vector of discriminant function coefficients (**a** in Display 7.3) being

$$\mathbf{a}' = [-0.0893, 0.156, 0.005, -0.177, -0.177]$$

and the threshold value being -30.363. The resulting classification rule becomes: classify to type I if $-0.0893 \times$ length $+ 0.15 \times$ Breadth $+ 0.005 \times$ fheight $-0.177 \times$ fbreadth > -30.363 and Type II otherwise. The same results can be obtained using the discrim function in S-PLUS:

```
dis<-discrim(Type~Length+Breadth+Height+Fheight+Fbreadth,
     data=Tibet,
family=Classical("homoscedastic"),prior="uniform")
dis
```

This gives the results shown in Table 7.2. The results given previously are found from Table 7.2 by simply subtracting the two sets of linear coefficients to give the vector of discriminant function coefficients and the two constants to give the threshold value:

```
const<-coef(dis)$constants
const[2]-const[1]
coefs<-coef(dis)$linear.coefficients
coefs[,1]-coefs[,2]
```

By loading the MASS library in both R and S-PLUS, Fisher's linear discriminant analysis can be applied using the lda function as

```
library(MASS)
dis<-lda(Type~Length+Breadth+Height+Fheight+Fbreadth,
     data=Tibet,prior=c(0.5,0.5))
```

Suppose now we have the following observations on two new skulls:

	Length	Breadth	Height	Fheight	Fbreadth
Skull 1:	171.0	140.5	127.0	69.5	137.0
Skull 2:	179.0	132.0	140.0	72.0	138.5

Table 7.2 Results from `discrim` on Tibetan Skull Data

	Length	Breadth	Height	Fheight	Fbreadth	N	Priors
Group means:							
1	174.82	139.35	132.00	69.824	130.35	17	0.5
2	185.73	138.73	134.77	76.467	137.50	15	0.5

Covariance Structure: homoscedastic

	Length	Breadth	Height	Fheight	Fbreadth
Length	59.013	9.008	17.219	20.120	20.110
Breadth		48.261	1.077	4.339	30.046
Height			36.198	4.838	4.108
Fheight				18.307	12.985
Fbreadth					43.696

Constants:

1	2
−514.9	−545.48

Linear Coefficients:

	X1	X2
Length	1.4683	1.5576
Breadth	2.3611	2.2053
Height	2.7522	2.7470
Fheight	0.7753	0.9525
Fbreadth	0.1948	0.3722

and wish to classify them to be type I or type II. We can calculate each skull's discriminant score as follows;

Skull 1: $-0.0893 \times 171.0 + 0.156 \times 140.5 + 0.005 \times 127.0 - 0.177 \times 69.5 - 0.177 \times 137.0 = -29.27$

Skull 2: $-0.893 \times 179.0 + 0.156 \times 132.0 + 0.005 \times 140.0 - 0.177 \times 72.0 - 0.177 \times 138.5 = -31.95$

Comparing each score to the threshold value of -30.363 leads to classifying skull 1 as type I and skull 2 as type II. We can use the `predict` function applied to the object `dis` to do the same thing:

```
newdata<-rbind(c(171,140.5,127.0,69.5,137.0),c(179.0,132.0,
    140.0,72.0,138.5))
dimnames(newdata)<-list(NULL,c("Length","Breadth","Height",
    "Fheight","Fbreadth"))
newdata<-data.frame(newdata)
predict(dis,newdata=newdata)
```

to give the following classification probabilities:

	Skull 1	Skull 2
Prob(Type I):	0.77695	0.22305
Prob(Type II):	0.19284	0.80716

Fisher's linear discriminant function is optimal when the data arise from populations having multivariate normal distributions with the same covariance matrices.

When the distributions are clearly non-normal an alternative approach is *logistic discrimination* (see, e.g., Anderson, 1972), although the results of both this and Fisher's method are likely to be very similar in most cases. When the two covariance matrices are thought to be unequal, then the linear discriminant function is no longer optimal and a quadratic version may be needed. Details are given in Everitt and Dunn (2001).

The quadratic discriminant function has the advantage of increased flexibility compared to the linear version. There is, however, a penalty involved in the form of potential overfitting, making the derived function poor at classifying new observations. Friedman (1989) attempts to find a compromise between the data variability of quadratic discrimination and the possible bias of linear discrimination by adopting a weighted sum of the two called *regularized discriminant analysis*.

7.2.3 Assessing the Performance of a Discriminant Function

How might we evaluate the performance of a discriminant function? One obvious approach would be to apply the function to the data from which it was derived and calculate the misclassification rate (this approach is known as the "plug-in" estimate). We can do this for the Tibetan skull data in R and S-PLUS by again using the `predict` function as follows:

```
group<-predict(dis,method="plug-in")$class
#in S-PLUS use predict(dis,method="plug-in")$group
table(group,Type)
```

leading to the following counts of correct and incorrect classifications:

	Correct group	
Allocated	1	2
1	14	3
2	3	12

The misclassification rate is 19%. This technique has the advantage of being extremely simple. Unfortunately, however, it generally provides a very poor estimate of the actual misclassification rate. In most cases the estimate obtained in this way will be highly optimistic. An improved estimate of the misclassification rate of a discriminant function may be obtained in a variety of ways (see Hand, 1998, for details). The most commonly used of the alternatives available is the so-called "leaving-one-out method," in which the discriminant function is derived from just $n - 1$ members of the sample and then used to classify the member not included. The process is carried out n times, leaving out each sample member in turn. We will illustrate the use of this approach later in the chapter.

7.3 More Than Two Groups: Multivariate Analysis of Variance (MANOVA) and Classification Functions

7.3.1 *Multivariate Analysis of Variance*

MANOVA is an extension of univariate analysis of variance procedures to multidimensional observations. Details of the technique for a one-way design are given in Display 7.3.

Display 7.3

Multivariate Analysis of Variance

- We assume we have multivariate observations for a number of individuals from m different populations where $m \geqslant 2$ and there are n_i observations from population i.
- The linear model for observation x_{ijk}, the jth observation on variable k in group i, $k = 1, \ldots, q$, $j = 1, \ldots, n_i$, $i = 1, \ldots, m$ is

$$x_{ijk} = \mu_k + \alpha_{ik} + \varepsilon_{ijk},$$

  where μ_k is a general effect for the kth variable, α_{ik} is the effect of group i on the kth variable, and ε_{ijk} is a random disturbance term.
- The vector $\varepsilon_{ij} = [\varepsilon_{ij1}, \ldots, \varepsilon_{ijq}]$ is assumed to have a multivariate normal distribution with null mean vector and covariance matrix, Σ, assumed to be the same in all m populations. The ε_{ij} of different individuals are assumed to be independent of one another.
- The hypothesis of equal mean vectors in the m populations can be written as

$$H_0: \alpha_{ik} = 0, \quad i = 1, \ldots, m, \quad k = 1, \ldots, q.$$

The multivariate analysis of variance is based on two matrices, **H** and **E**, the elements of which are defined as follows:

$$h_{rs} = \sum_{i=1}^{k} n_i (\bar{x}_{ir} - \bar{x}_r)(\bar{x}_{is} - \bar{x}_s), \quad r, s = 1, \ldots, q,$$

$$e_{rs} = \sum_{i=1}^{k} \sum_{j=1}^{n_i} (\bar{x}_{ijr} - \bar{x}_{ir})(\bar{x}_{ijs} - \bar{x}_{is}), \quad r, s = 1, \ldots, q,$$

where $\bar{x}_{ir}$ is the mean of variable r in group i, and $\bar{x}_r$ is the grand mean of variable r.
- The diagonal elements of **H** and **E** are, respectively, the between-groups sum of squares for each variable, and the within-group sum of squares for the variable.

- The off-diagonal elements of **H** and **E** are the corresponding sums of cross-products for pairs of variables.
- In the multivariate situation when $m > 2$ there is no single test statistic that is always the most powerful for detecting all types of departures from the null hypothesis of the mean vectors of the populations.
- A number of different test statistics have been proposed that may lead to different conclusions when used in the same data set, although on most occasions they will not.
- The following are the principal test statistics for the multivariate analysis of variance

  (a) *Wilks' determinantal ratio*

  $$\Lambda = \frac{|\mathbf{E}|}{|\mathbf{H} + \mathbf{E}|}$$

  (b) *Roy's greatest root*
  Here the criterion is the largest eigenvalue of $\mathbf{E}^{-1}\mathbf{H}$
  (c) *Lawley–Hotelling trace*

  $$t = \text{trace}(\mathbf{E}^{-1}\mathbf{H}).$$

  (d) *Pillai trace*
  $$v = \text{trace}[\mathbf{H}(\mathbf{H} + \mathbf{E})^{-1}].$$

- Each test statistic can be converted into an approximate F-statistic that allows associated p-values to be calculated. For details see Tabachnick and Fidell (2000).
- When there are only two groups all four test criteria above are equivalent and lead to the same F value as Hotelling's T^2 as given in Display 7.1.

We will illustrate the application of MANOVA using the data on skull measurements in different epochs met in Chapter 5 (see Table 5.8). We can apply a one-way MANOVA to these data and get values for each of the four test statistics described in Display 7.3 using the following R and S-PLUS code:

R

```
attach(skulls)
skulls.manova < −manova(cbind(MB, BH, BL, NH)~EPOCH)
summary(skulls.manova, test = "Pillai")
summary(skulls.manova, test = "Wilks")
summary(skulls.manova, test = "Hotelling")
summary(skulls.manova, test = "Roy")
```

S-PLUS

```
attach(skulls)
skulls.manova < —manova(cbind(MB, BH, BL, NH)~EPOCH)
summary(skulls.manova, test = "pillai")
summary(skulls.manova, test = "wilks")
summary(skulls.manova, test = "hotelling − lawley")
summary(skulls.manova, test = "roylargest")
```

The results are shown in Table 7.3. There is very strong evidence that the mean vectors of the five epochs differ.

The tests applied in MANOVA assume multivariate normality for the error terms in the corresponding model (see Display 7.3). An informal assessment of this assumption can be made using the chi-square plot described in Chapter 1 applied to the residuals from fitting the one-way MANOVA model; note that the residuals in this case are each four-dimensional vectors. The required R and S-PLUS instruction is

```
chisplot(residuals(skulls.manova))
```

This gives the diagram shown in Figure 7.1. There is no evidence of a departure from multivariate normality.

7.3.2 Classification Functions and Canonical Variates

When there is an interest in classification of multivariate observations where there are more than two groups, a series of classification functions can be derived based on the two-group approach described in Section 7.2. Details are given in Display 7.4 for the three group situation.

Table 7.3 Multivariate Analysis of Variance Results for Egyptian Skull Data

```
        Df Pillai Trace approx. F num df den df P-value
   EPOCH 4           0.35       3.51      16      580        0
Residuals 145
        Df Wilks Lambda approx. F num df den df P-value
   EPOCH 4           0.66       3.9       16     434.45      0
Residuals 145
        Df Hotelling-Lawley approx. F num df den df P-value
   EPOCH 4           0.48       4.23      16      562        0
Residuals 145
        Df Roy Largest approx. F num df den df P-value
   EPOCH 4           0.43      15.41       4      145        0
Residuals 145
```

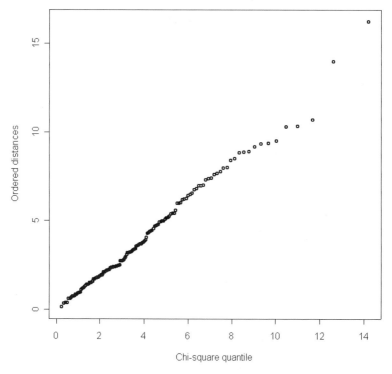

Figure 7.1 Chi-square plot of residuals from fitting one-way MANOVA model to Egyptian skull data.

Display 7.4
Classification Functions for Three Groups

- When more than two groups are involved, the rule for allocating to two multivariate normal distributions with the same covariance matrix can be applied to each pair of groups in turn to derive a series of classification functions.
- For three groups, for example, the sample versions of the functions would be:

$$h_{12}(\mathbf{x}) = (\bar{\mathbf{x}}_1 - \bar{\mathbf{x}}_2)'\mathbf{S}^{-1}\left[\mathbf{x} - \frac{1}{2}(\bar{\mathbf{x}}_1 + \bar{\mathbf{x}}_2)\right],$$

$$h_{13}(\mathbf{x}) = (\bar{\mathbf{x}}_1 - \bar{\mathbf{x}}_3)'\mathbf{S}^{-1}\left[\mathbf{x} - \frac{1}{2}(\bar{\mathbf{x}}_1 + \bar{\mathbf{x}}_3)\right],$$

$$h_{23}(\mathbf{x}) = (\bar{\mathbf{x}}_2 - \bar{\mathbf{x}}_3)'\mathbf{S}^{-1}\left[\mathbf{x} - \frac{1}{2}(\bar{\mathbf{x}}_2 + \bar{\mathbf{x}}_3)\right],$$

where $\mathbf{S}$ is the pooled within-groups covariance matrix calculated over *all* three groups.

• The classification rule now becomes:

Allocate to G_1 if $h_{12}(x) > 0$ and $h_{13}(x) > 0$;
Allocate to G_2 if $h_{12}(x) < 0$ and $h_{23}(x) > 0$;
Allocate to G_3 if $h_{13}(x) < 0$ and $h_{23}(x) < 0$.

The classification functions allow observations to be classified optimally, but interest may also lie in identifying the dimensions of the multivariate space of the observed variables that are of most importance in distinguishing between the groups. For two groups the single dimension is given by Fisher's linear discriminant function, which, as described in Display 7.3, arises as the single solution of the equation

$$(\mathbf{B} - \lambda\mathbf{S})\mathbf{a} = 0.$$

When there are more than two groups however, this equation will have more then one solution, reflecting the fact that more than one direction is needed to describe the differences between the mean vectors of the groups. With g groups and q variables there will be $\min(q, g - 1)$ solutions. These best separating dimensions are known as *canonical variates*. We can find them and the relevant classification for the Egyptian skull data by again using the discrim function (or alternatively the lda function although the options and output are not quite so comprehensive):

```
dis<-discrim(EPOCH~MB+BH+BL+NH,data=skulls,family=Canonical
    ("homoscedastic"))
dis
summary(dis)
```

An edited version of the results is shown in Table 7.4. To form the classification functions described in Display 7.4 we need to look at the "linear coefficients" and the "constants" in this table. For example, the classification function for epochs, c4000BC, and c3300BC, can be found as

```
const <- coef(dis)$constants
t12<-const[2] - const[1]
coefs <- coef(dis)$linear.coefficients
h12<-coefs[, 1] - coefs[, 2]
```

This gives the necessary vector of constants and threshold to form the first of the required classification functions; similarly, the remaining classification functions, h13,h14,...,h45, and thresholds, t12,t13,...,t45, can be found. They may then be applied to the four measurements on a new skull as indicated by the rule in Display 7.4 extended in an obvious way to the five-group situation, to classify the skull into one of the five epochs (see Exercise 7.4).

Table 7.4 Edited Results from Applying the `discrim` Function to the Egyptian Skull Data

```
Group means:
            MB       BH       BL       NH    N   Priors
c4000BC  131.37   133.60   99.167   50.533   30   0.2
c3300BC  132.37   132.70   99.067   50.233   30   0.2
c1850BC  134.47   133.80   96.033   50.567   30   0.2
 c200BC  135.50   132.30   94.533   51.967   30   0.2
 cAD150  136.17   130.33   93.500   51.367   30   0.2

Covariance Structure: homoscedastic
          MB       BH       BL       NH
MB    21.111    0.037    0.079    2.009
BH              23.485    5.200    2.845
BL                       24.179    1.133
NH                                10.153

Canonical Coefficients:
          dim1         dim2         dim3         dim4
MB     0.126676     0.038738    -0.092768    -0.14883986
BH    -0.037032     0.210098     0.024568     0.00042008
BL    -0.145125    -0.068114    -0.014749    -0.13250077
NH     0.082851    -0.077293     0.294589    -0.06685888

Singular Values:
  dim1     dim2      dim3      dim4
3.9255    1.189    0.75451    0.27062

Constants:
c4000BC   c3300BC   c1850BC    c200BC    cAD150
-914.53   -915.35   -925.34   -923.42   -914.12

Linear Coefficients:
       c4000BC   c3300BC   c1850BC    c200BC   cAD150
MB      6.0012    6.0515    6.1507    6.1850   6.2209
BH      4.7674    4.7316    4.8090    4.7380   4.6650
BL      2.9569    2.9617    2.8189    2.7647   2.7395
NH      2.1238    2.0938    2.1013    2.2583   2.2154

Plug-in classification table:
          c4000BC   c3300BC   c1850BC   c200BC   cAD150    Error   Posterior.Error
c4000BC      12        8         4        4        2     0.60000          0.60390
c3300BC      10        8         5        4        3     0.73333          0.68144
c1850BC       4        4        15        2        5     0.50000          0.62486
 c200BC       3        3         7        5       12     0.83333          0.73753
 cAD150       2        4         4        9       11     0.63333          0.51736
Overall                                                 0.66000          0.63302
(from=rows,to=columns)

Rule Mean Square Error: 0.71928
(conditioned on the training data)

Cross-validation table:
          c4000BC   c3300BC   c1850BC   c200BC   cAD150    Error   Posterior.Error
c4000BC       9       10         5        4        2     0.70000          0.61844
c3300BC      11        7         5        4        3     0.76667          0.65822
c1850BC       6        4        12        2        6     0.60000          0.64549
 c200BC       3        3         7        5       12     0.83333          0.72218
 cAD150       2        4         4       10       10     0.66667          0.52258
Overall                                                 0.71333          0.63338
(from=rows,to=columns)
```

The coefficients defining the canonical variates are to found under "canonical coefficients" in Table 7.4. Here, with five groups and four variables, there are four such variates. To see how the canonical variates discriminate between the groups, it is often useful to plot the group canonical variate means. For example, we can plot the means for the first two canonical variates as

```
dsfs1<-c(0.13,-0.04,-0.15,0.08)%*%t(skulls[,-1])
dsfs2<-c(0.04,0.21,-0.068,-0.08)%*%t(skulls[,-1])
m1<-
    c(mean(dsfs1[1:30]),mean(dsfs1[31:60]),mean(dsfs1[61:90]),
    mean(dsfs1[91:120]),mean(dsfs1[121:150]))
m2<-
    c(mean(dsfs2[1:30]),mean(dsfs2[31:60]),mean(dsfs2[61:90]),
    mean(dsfs2[91:120]),mean(dsfs2[121:150]))
plot(m1,m2,type="n",xlab="CV1",ylab="CV2",xlim=c(0.5,3))
text(m1,m2,labels=c("c4000BC","c3300BC","c1850BC","c200BC",
    "cAD150"))
```

The result is shown in Figure 7.2. The first canonical variate separates the two earliest epochs from the other three and the second separates c1850BC from the remaining four.

The "plug-in" estimate of the misclassification rate is shown in Table 7.4. Also shown is the more realistic "leave-out-one" or "cross-validation" estimate. There is a considerable amount of misclassification particularly for c200BC and cAD150.

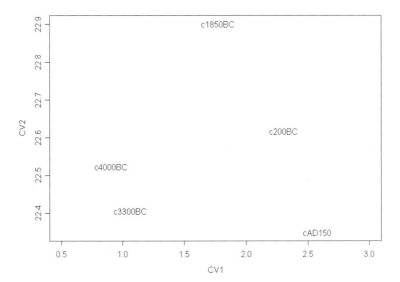

Figure 7.2 Epoch means for the first two canonical variates.

Table 7.5 SIDS data

Group	HR	BW	Factor68	Gesage
1	115.6	3060	0.291	39
1	108.2	3570	0.277	40
1	114.2	3950	0.390	41
1	118.8	3480	0.339	40
1	76.9	3370	0.248	39
1	132.6	3260	0.342	40
1	107.7	4420	0.310	42
1	118.2	3560	0.220	40
1	126.6	3290	0.233	38
1	138.0	3010	0.309	40
1	127.0	3180	0.355	40
1	127.7	3950	0.309	40
1	106.8	3400	0.250	40
1	142.1	2410	0.368	38
1	91.5	2890	0.223	42
1	151.1	4030	0.364	40
1	127.1	3770	0.335	42
1	134.3	2680	0.356	40
1	114.9	3370	0.374	41
1	118.1	3370	0.152	40
1	122.0	3270	0.356	40
1	167.0	3520	0.394	41
1	107.9	3340	0.250	41
1	134.6	3940	0.422	41
1	137.7	3350	0.409	40
1	112.8	3350	0.241	39
1	131.3	3000	0.312	40
1	132.7	3960	0.196	40
1	148.1	3490	0.266	40
1	118.9	2640	0.310	39
1	133.7	3630	0.351	40
1	141.0	2680	0.420	38
1	134.1	3580	0.366	40
1	135.5	3800	0.503	39
1	148.6	3350	0.272	40
1	147.9	3030	0.291	40

(*Continued*)

Table 7.6 (*Continued*)

Group	HR	BW	Factor68	Gesage
1	162.0	3940	0.308	42
1	146.8	4080	0.235	40
1	131.7	3520	0.287	40
1	149.0	3630	0.456	40
1	114.1	3290	0.284	40
1	129.2	3180	0.239	40
1	144.2	3580	0.191	40
1	148.1	3060	0.334	40
1	108.2	3000	0.321	37
1	131.1	4310	0.450	40
1	129.7	3975	0.244	40
1	142.0	3000	0.173	40
1	145.5	3940	0.304	41
2	139.7	3740	0.409	40
2	121.3	3005	0.626	38
2	131.4	4790	0.383	40
2	152.8	1890	0.432	38
2	125.6	2920	0.347	40
2	139.5	2810	0.493	39
2	117.2	3490	0.521	38
2	131.5	3030	0.343	37
2	137.3	2000	0.359	41
2	140.9	3770	0.349	40
2	139.5	2350	0.279	40
2	128.4	2780	0.409	39
2	154.2	2980	0.388	40
2	140.7	2120	0.372	38
2	105.5	2700	0.314	39
2	121.7	3060	0.405	41

7.4 Summary

Grouped multivariate data occur frequently in practice. The appropriate method of analysis depends on the question of most interest to the investigator. Hotelling's T^2 and MANOVA are used to assess formal hypothesis about population mean vectors. Where there is evidence of a difference then the construction of a classification rule is often (but not always) of interest. A range of other discriminant procedures are available in the MASS library, and readers are encouraged to investigate.

Exercises

7.1   In a two-group discriminant situation, if members of one group have a y-value of -1 and those of the other group a value of 1, show that the coefficients in a regression of y on $x_1, x_2, \ldots, x_q$ are proportional to $\mathbf{S}^{-1}(\bar{\mathbf{x}}_1 - \bar{\mathbf{x}}_2)$, the coefficients of Fisher's linear discriminant function.

7.2   In the two-group discrimination problem, suppose that

$$f_i(x) = \binom{n}{x} p_i^x (1 - p_i)^{n-x}, \qquad 0 < p_i < 1, \quad i = 1, 2,$$

where p_1 and p_2 are known. If π_1 and π_2 are the prior probabilities of the two groups, devise the classification rule using the approach described in Display 7.3.

7.3   The data shown in Table 7.5 were collected by Spicer et al. (1987) in an investigation of sudden infant death syndrome (SIDS). The two groups here consist of 16 SIDS victims and 49 controls. The Factor68 variable arises from spectral analysis of 24 hour recordings of electrocardiograms and respiratory movements made on each child. All the infants have a gestational age of 37 weeks or more and were regarded as full term.

   (i)   Construct Fisher's linear discriminant function using only the Factor68 and Birthweight variables. Show the derived discriminant function on a scatterplot of the data.
   (ii) Construct the discriminant function based on all four variables and find an appropriate estimate of the misclassification rate.
   (iii) How would you incorporate prior probabilities into your discriminant function?

7.4   Find all the classification functions for the Egyptian skull data and use them to allocate a new skull with the following measurements:

   MB: 133.0
   BH: 130.0
   BL: 95.0
   NH: 50.0

8
Multiple Regression and Canonical Correlation

8.1 Introduction

In this chapter we discuss two related but separate techniques, *multiple regression* and *canonical correlation*. The first of these is not strictly a multivariate procedure; the reasons for including it in this book are that it provides some useful basic material both for the discussion of canonical correlation in this chapter and modelling longitudinal data in Chapter 9.

8.2 Multiple Regression

Multiple linear regression represents a generalization, to more than a single explanatory variable, of the simple linear regression model met in all introductory statistics courses. The method is used to investigate the relationship between a dependent variable, y, and a number of explanatory variables $x_1, x_2, \ldots, x_q$. Details of the model, including the estimation of its parameters by least squares and the calculation of standard errors are given in Display 8.1. Note in particular that the explanatory variables are, strictly, not regarded as random variables at all so that multiple regression is essentially a *univariate* technique with the only random variable involved being the response, y. Often the technique is referred to as being *multivariable* to properly distinguish it from genuinely multivariate procedures.

As an example of the application of multiple regression we can apply it to the air pollution data introduced in Chapter 3 (see Table 3.1), with SO_2 level as the dependent variable and the remaining variables being explanatory. The model can be applied in R and S-PLUS® and the results summarized using

```
attach(usair.dat)
usair.fit<-lm(SO2~Neg.Temp + Manuf + Pop + Wind +
    Precip + Days)
summary(usair.fit)
```

Display 8.1

Multiple Regression Model

- The multiple linear regression model for a response variable y with observed values $y_1, y_2, \ldots, y_n$ and q explanatory variables, $x_1, x_2, \ldots, x_q$, with observed values $x_{i1}, x_{i2}, \ldots, x_{iq}$ for $i = 1, 2, \ldots, n$, is

$$y_i = \beta_0 + \beta_1 x_{i1} + \beta_2 x_{i2} + \cdots + \beta_q x_{iq} + \varepsilon_i.$$

- The regression coefficients $\beta_1, \beta_2, \ldots, \beta_q$ give the amount of change in the response variable associated with a unit change in the corresponding explanatory variable, *conditional* on the other explanatory variables in the model remaining unchanged.
- The explanatory variables are strictly assumed to be fixed; that is, they are not random variables. In practice, where this is rarely the case, the results from a multiple regression analysis are interpreted as being *conditional* on the observed values of the explanatory variables.
- The residual terms in the model, $\varepsilon_i, i = 1, \ldots, n$, are assumed to have a normal distribution with mean zero and variance σ^2. This implies that, for given values of the explanatory variables, the response variable is normally distributed with a mean that is a linear function of the explanatory variables and a variance that is not dependent on these variables. Consequently an equivalent way of writing the multiple regression model is as $y \sim N(\mu, \sigma^2)$ where $\mu = \beta_0 + \beta_1 x_1 + \cdots + \beta_q x_q$.
- The "linear" in multiple linear regression refers to the parameters rather than the explanatory variables, so the model remains linear if, for example, a quadratic term for one of these variables is included. (An example of a *nonlinear model* is $y = \beta_1 e^{\beta_2 x_{i1}} + \beta_3 e^{\beta_4 x_{i2}} + \varepsilon_i$.)
- The aim of multiple regression is to arrive at a set of values for the regression coefficients that makes the values of the response variable predicted from the model as close as possible to the observed values.
- The least-squares procedure is used to estimate the parameters in the multiple regression model.
- The resulting estimators are most conveniently written with the help of some matrices and vectors. By introducing a vector $\mathbf{y}' = [y_1, y_2, \ldots, y_n]$ and an $n \times (q + 1)$ matrix $\mathbf{X}$ given by

$$\mathbf{X} = \begin{pmatrix} 1 & x_{11} & x_{12} & \cdots & x_{1q} \\ 1 & x_{21} & x_{22} & \cdots & x_{2q} \\ \vdots & \vdots & \vdots & \vdots & \vdots \\ 1 & x_{n1} & x_{n2} & \cdots & x_{nq} \end{pmatrix},$$

we can write the multiple regression model for the n observations concisely as

$$\mathbf{y} = \mathbf{X}\boldsymbol{\beta} + \boldsymbol{\varepsilon},$$

where $\boldsymbol{\varepsilon}' = [\varepsilon_1, \varepsilon_2, \ldots, \varepsilon_n]$ and $\boldsymbol{\beta}' = [\beta_0, \beta_1, \ldots, \beta_q]$.

- The least-squares estimators of the parameters in the multiple regression model are given by the set of equations

$$\hat{\beta} = (\mathbf{X}'\mathbf{X})^{-1}\mathbf{X}'\mathbf{y}.$$

- More details of the least-squares estimation process are given in Rawlings et al. (1998).
- The variation in the response variable can be partitioned into a part due to regression on the explanatory variables and a residual as for simple linear regression. The can be arranged in an analysis of variance table as follows:

Source	DF	SS	MS	F
Regression	q	RGSS	RGSS/q	RGMS/RSMS
Residual	$n - q - 1$	RSS	RSS/$n - q - 1$	

- The residual mean square s^2 is an estimator of σ^2.
- The covariance matrix of the parameter estimates in the multiple regression model is estimated from

$$\mathbf{S}_{\hat{\beta}} = s^2(\mathbf{X}'\mathbf{X})^{-1}.$$

The diagonal elements of this matrix give the variances of the estimated regression coefficients and the off-diagonal elements their covariances.

- A measure of the fit of the model is provided by the *multiple correlation coefficient*, R, defined as the correlation between the observed values of the response variable, y_1, K, y_n, and the values predicted by the fitted model, that is,

$$\hat{y}_i = \hat{\beta}_0 + \hat{\beta}_1 x_{i1} + \cdots + \hat{\beta}_q x_{iq}$$

- The value of R^2 gives the proportion of variability in the response variable accounted for by the explanatory variables.

The results are shown in Table 8.1. The F statistic for testing the hypothesis that all six regression coefficients in the model are zero is 11.48 with 6 and 34 degrees of freedom. The associated p-value is very small and the hypothesis should clearly be rejected. The t-statistics suggest that *Manuf* and *Pop* are the most important predictors of sulphur dioxide level. The square of the multiple correlation coefficient is 0.67 showing that 67% of the variation in SO_2 level is accounted for by the six explanatory variables.

When applying multiple regression in practice, of course, analysis would continue to try to identify a more parsimonious model, followed by examination of residuals from the final model to check assumptions. We shall not do this since

Table 8.1 Results of Multiple Regression Applied to Air Pollution Data

Covariate	Estimated regression coefficient	Standard error	t-value	p
(Intercept)	111.7285	47.3181	2.3612	0.0241
Neg. temp	1.2679	0.6212	2.0412	0.0491
Manuf	0.0649	0.0159	4.1222	0.0002
Pop	−0.0393	0.0151	−2.5955	0.0138
Wind	−3.1814	1.8150	−1.7528	0.0887
Precip	0.5124	0.3628	1.4124	0.1669
Days	−0.0521	0.1620	−0.3213	0.7500

Residual standard error: 14.64 on 34 degrees of freedom.
Multiple R-squared: 0.6695.
F-statistic: 11.48 on 6 and 34 degrees of freedom, the p-value is 5.419e − 007.

here we are largely interested in the univariate multiple regression model merely as a convenient stepping stone to discuss a number of multivariate procedures beginning with canonical correlation analysis.

8.3 Canonical Correlations

Multiple regression is concerned with the relationship between a single variable y and a set of variables $x_1, x_2, \ldots, x_q$. Canonical correlation analysis extends this idea to investigating the relationship between two sets of variables, *each* containing more than a single member. For example, in psychology an investigator may measure a set of aptitude variables and a set of achievement variables on a sample of students. In marketing, a similar example might involve a set of price indices and a set of prediction indices. The objective of canonical correlation analysis is to find linear functions of one set of variables that maximally correlate with linear functions of the other set of variables. In many circumstances one set will contain multiple dependent variables and the other multiple independent or explanatory variables and then canonical correlation analysis might be seen as a way of predicting multiple dependent variables from multiple independent variables. Extraction of the coefficients that define the required linear functions has similarities to the process of finding principal components as described in Chapter 3. Some of the steps are described in Display 8.2.

To begin we shall illustrate the application of canonical correlation analysis on a data set reported over 80 years ago by Frets (1921). The data are given in Table 8.2 and give head measurements (in millimeters) for each of the first two adult sons in 25 families. Here the family is the "individual" in our data set and the four head measurements are the variables. The question that was of interest to Frets was whether there is a relationship between the head measurements for pairs of sons? We shall address this question by using canonical correlation analysis.

Here we shall develop the canonical correlation analysis from first principles as detailed in Display 8.2. Assuming the head measurements data are contained in the

Display 8.2
Canonical Correlation Analysis (CCA)

- The purpose of canonical correlation analysis is to characterize the independent statistical relationships that exist between two sets of variables, $\mathbf{x}' = [x_1, x_2, \ldots, x_{q_1}]$ and $\mathbf{y}' = [y_1, y_2, \ldots, y_{q_2}]$.
- The overall $(q_1 + q_2) \times (q_1 + q_2)$ correlation matrix contains all the information on associations between pairs of variables in the two sets, but attempting to extract from this matrix some idea of the association between the two sets of variables is not straightforward. This is because the correlations between the two sets may not have a consistent pattern; and these between-set correlations need to be adjusted in some way for the within-set correlations.
- The question is "How do we quantify the association between the two sets of variables $\mathbf{x}$ and $\mathbf{y}$?"
- The approach adopted in CCA is to take the association between $\mathbf{x}$ and $\mathbf{y}$ to be the largest correlation between two single variables u_1 and v_1 derived from $\mathbf{x}$ and $\mathbf{y}$, with u_1 being a linear combination of $x_1, x_2, \ldots, x_{q_1}$ and v_1 being a linear combination of $y_1, y_2, \ldots, y_{q_2}$.
- But often a single pair of variables, (u_1, v_1) is not sufficient to quantify the association between the x and y variables, and we may need to consider some or all of s pairs $(u_1, v_1), (u_2, v_2), \ldots, (u_s, v_s)$ to do this, where $s = \min(q_1, q_2)$.
- Each u_i is a linear combination of the variables in $\mathbf{x}$, $u_i = \mathbf{a}_i'\mathbf{x}$, and each v_i is a linear combination of the variables $\mathbf{y}$, $v_i = \mathbf{b}_i\mathbf{y}$, with the coefficients $(\mathbf{a}_i, \mathbf{b}_i)(i = 1, \ldots, s)$ being chosen so that the u_i and v_i satisfy the following:

  (1) The u_i are mutually uncorrelated, i.e., $\mathrm{cov}(u_i, u_j) = 0$ for $i \neq j$.
  (2) The v_i are mutually uncorrelated, i.e., $\mathrm{cov}(v_i, v_j) = 0$ for $i \neq j$.
  (3) The correlation between u_i and v_i is R_i for $i = 1, \ldots, s$, where $R_1 > R_2 > \cdots > R_s$.
  (4) The u_i are uncorrelated with all v_j except v_i, i.e., $\mathrm{cov}(u_i, v_j) = 0$ for $i \neq j$.

- The linear combinations u_i and v_i are often referred to as *canonical variates*, a name used previously in Chapter 7 in the context of multiple discriminant function analysis. In fact there is a link between the two techniques. If we perform a canonical correlation analysis with the data $\mathbf{X}$ defining one set of variables and a matrix of group indicators, $\mathbf{G}$, as the other we obtain the linear discriminant functions. Details are given in Mardia et al. (1979).
- The vectors $\mathbf{a}_i$ and $\mathbf{b}_i$ $i = 1, \ldots, s$, which define the required linear combinations of the x and y variables, are found as the eigenvectors of matrices $\mathbf{E}_1(q_1 \times q_1)$ (the $\mathbf{a}_i$) and $\mathbf{E}_2(q_2 \times q_2)$ (the $\mathbf{b}_i$) defined as

$$\mathbf{E}_1 = \mathbf{R}_{11}^{-1}\mathbf{R}_{12}\mathbf{R}_{22}^{-1}\mathbf{R}_{21}, \qquad \mathbf{E}_2 = \mathbf{R}_{22}^{-1}\mathbf{R}_{21}\mathbf{R}_{11}^{-1}\mathbf{R}_{12},$$

where $\mathbf{R}_{11}$ is the correlation matrix of the variables in $\mathbf{x}$, $\mathbf{R}_{22}$ is the correlation matrix of the variables in $\mathbf{y}$, and $\mathbf{R}_{12}(=\mathbf{R}_{21})$ the $q_1 \times q_2$ matrix of correlations across the two sets of variables.

- The canonical correlations $R_1, R_2, \ldots, R_s$ are obtained as the square roots of the nonzero eigenvalues of either $\mathbf{E}_1$ or $\mathbf{E}_2$.
- The s canonical correlations $R_1, R_2, \ldots, R_s$ express the association between the $\mathbf{x}$ and $\mathbf{y}$ variables after removal of the within-set correlation.
- More details of the calculations involved and the theory behind canonical correlation analysis are given in Krzanowski (1988).
- Inspection of the coefficients of each original variable in each canonical variate can provide an interpretation of the canonical variate in much the same way as interpreting principal components (see Chapter 3). Such interpretation of the canonical variates may help to describe just how the two sets of original variables are related (see Krzanowski 2004).
- In practice, interpretation of canonical variates can be difficult because of the possibly very different variances and covariances among the original variables in the two sets, which affects the sizes of the coefficients in the canonical variates. Unfortunately there is no convenient normalization to place all coefficients on an equal footing (see Krzanowski, 2004).
- In part this problem can be dealt with by restricting interpretation to the standardized coefficients, that is, the coefficients that are appropriate when the original variables have been standardized.

data frame, `headsize`, the necessary R and S-PLUS code is:

```
headsize.std<-sweep(headsize,2,
  sqrt(apply(headsize,2,var)),FUN="/")
#standardize head measurements by
#dividing by the appropriate standard deviation
#
#
headsize1<-headsize.std[,1:2]
headsize2<-headsize.std[,3:4]
#
#find all the matrices necessary for calculating the
#canonical variates and canonical correlations
#
R11<-cor(headsize1)
R22<-cor(headsize2)
R12<-c(cor(headsize1[,1],headsize2[,1]),cor(headsize1[,1],
  headsize2[,2]),
cor(headsize1[,2],headsize2[,1]),cor(headsize1[,2],
  headsize2[,2]))
```

Table 8.2 Head Sizes in Pairs of Sons (mm)

x_1	x_2	x_3	x_4
191	155	179	145
195	149	201	152
181	148	185	149
183	153	188	149
176	144	171	142
208	157	192	152
189	150	190	149
197	159	189	152
188	152	197	159
192	150	187	151
179	158	186	148
183	147	174	147
174	150	185	152
190	159	195	157
188	151	187	158
163	137	161	130
195	155	183	158
186	153	173	148
181	145	182	146
175	140	165	137
192	154	185	152
174	143	178	147
176	139	176	143
197	167	200	158
190	163	187	150

x_1 = head length of first son; x_2 = head breadth of first son; x_3 = head length of second son; x_4 = head breadth of second son.

```
#
R12<-matrix(R12,ncol=2,byrow=T)
R21<-t(R12)
#
#see display 8.2 for relevant equations
E1<-solve(R11)%*%R12%*%solve(R22)%*%R21
E2<-solve(R22)%*%R21%*%solve(R11)%*%R12
#
E1
E2
#
eigen(E1)
eigen(E2)
```

The results are shown in Table 8.3. Here the four linear functions are found to be

$$u_1 = 0.69x_1 + 0.72x_2, \quad v_1 = 0.74x_1 + 0.67x_2,$$
$$u_2 = 0.71x_1 - 0.71x_2, \quad v_2 = 0.70x_1 - 0.71x_2.$$

Table 8.3 Canonical Correlation Analysis Results on Headsize Data

$$\mathbf{E}_1 = \begin{bmatrix} 0.306 & 0.305 \\ 0.314 & 0.319 \end{bmatrix}$$

$$\mathbf{E}_2 = \begin{bmatrix} 0.330 & 0.324 \\ 0.295 & 0.295 \end{bmatrix}$$

Eigenvalues of $\mathbf{E}_1$ and $\mathbf{E}_2$ are 0.62 and 0.0029, giving the canonical correlations as $\sqrt{0.6215} = 0.7885$ and $\sqrt{0.0029} = 0.0537$ The respective eigenvectors are;

$$\mathbf{a}_1' = [0.695, 0.719],$$
$$\mathbf{a}_2' = [0.709, -0.705],$$
$$\mathbf{b}_1' = [0.742, 0.670],$$
$$\mathbf{b}_2' = [0.705, 0.711].$$

The first canonical variate for both first and second sons is simply a weighted sum of the two head measurements and might be labelled "girth"; these two variates have a correlation of 0.79. Each second canonical variate is a weighted difference of the two head measurements and can be interpreted roughly as head "shape"; here the correlation is 0.05. (Girth and shape are defined to be uncorrelated within first and second sons, and also between first and second sons.)

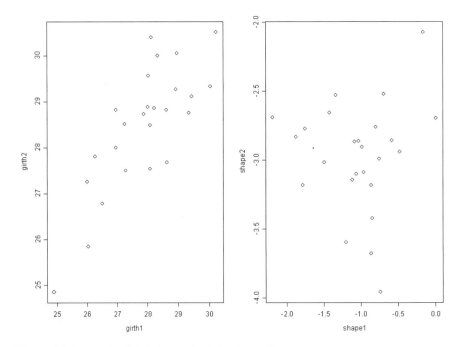

Figure 8.1 Scatterplots of girth and shape for first and second sons.

In this example it is clear that the association between the two head measurements of first and second sons is almost entirely expressed through the "girth" variables with the two "shape" variables being almost uncorrelated. The association between the two sets of measurements is essentially one-dimensional. A scatterplot of girth for first and second sons and a similar plot for shape reinforce this conclusion. The plots are both shown in Figure 8.1 which is obtained as follows;

```
girth1<-0.69*headsize.std[,1]+0.72*headsize.std[,2]
girth2<-0.74*headsize.std[,3]+0.67*headsize.std[,4]
shape1<-0.71*headsize.std[,1]-0.71*headsize.std[,2]
shape2<-0.70*headsize.std[,3]-0.71*headsize.std[,4]
#
cor(girth1,girth2)
cor(shape1,shape2)
#
par(mfrow=c(1,2))
plot(girth1,girth2)
plot(shape1,shape2)
```

The correlations between girth for first and second sons and similarly for shape calculated by this code are included to show that they give the same values (apart from rounding differences) as the canonical correlation analysis.

We can now move on to a more substantial example taken from Afifi et al. (2004), and also discussed by Krazanowski (2004). The data for this example arise from a study of depression amongst 294 respondents in Los Angeles. The two sets of variables of interest were "health variables," namely the CESD (the sum of 20 separate numerical scales measuring different aspects of depression) and a measure of general health and "personal" variables, of which there were four, gender, age, income and educational level (numerically, coded from the lowest "less than high school," to the highest, "finished doctorate"). The sample correlation matrix between these variables is given in Table 8.4. Here the maximum number of canonical variate pairs is 2, and they can be found using the following R and S-PLUS code:

```
r22<-matrix(c(1.0,0.044,-0.106,-0.180,0.044,1.0,-0.208,
    -0.192,-0.106,-0.208,1.0,0.492,-0.180,-0.192,0.492,1.0),
    ncol=4,byrow=T)
r11<-matrix(c(1.0,0.212,0.212,1.0),ncol=2,byrow=2)
```

Table 8.4 Sample Correlation Matrix for the Six Variables in the Los Angeles Depression Study

	CESD	Health	Gender	Age	Education	Income
CESD	1.0	0.121	0.124	−0.164	−0.101	−0.158
Health	0.212	1.0	0.098	0.308	−0.270	−0.183
Gender	0.124	0.098	1.0	0.044	−0.106	−0.180
Age	−0.164	0.308	0.044	1.0	−0.208	−0.192
Education	−0.101	−0.270	−0.106	−0.208	1.0	0.492
Income	−0.158	−0.183	−0.180	−0.192	0.492	1.0

```
r12<-matrix(c(0.124,-0.164,-0.101,-0.158,0.098,0.308,
   -0.270,-0.183),ncol=4,byrow=T)
r21<-t(r12)
#
E1<-solve(r11)%*%r12%*%solve(r22)%*%r21
E2<-solve(r22)%*%r21%*%solve(r11)%*%r12
#
E1
E2
#
eigen(E1)
eigen(E2)
```

The results are shown in Table 8.5. The first canonical correlation is 0.409 which if tested as outlined in Exercise 8.3 and has an associated p-value that is very small. There is strong evidence that the first canonical correlation is significant. The corresponding variates, in terms of standardized original variables, are

$$u_1 = 0.461 \text{ CESD} - 0.900 \text{ Health},$$

$$v_1 = 0.024 \text{ Gender} + 0.885 \text{ Age} - 0.402 \text{ Education} + 0.126 \text{ Income}.$$

High coefficients correspond to CESD (positively) and health (negatively) for the perceived health variables, and to age (positively) and education (negatively)

Table 8.5 Canonical Variates and Correlation for Los Angeles Depression Study Variables

```
eigen(R1)
$values:
[1] 0.16763669 0.06806171

$vectors:
numeric matrix: 2 rows, 2 columns.
             [,1]        [,2]
[1,]   0.4610975 -0.9476307
[2,] -0.8998655 -0.3193681

eigen(R2)
$values:
[1] 1.676367e-001 6.806171e-002 -1.734723e-018 0.000000e+000

$vectors:
numeric matrix: 4 rows, 4 columns.
              [,1]        [,2]        [,3]        [,4]
[1,]   0.02424121  0.6197600 -0.03291919 -0.9378101
[2,]   0.88498865 -0.6301703 -0.16889507 -0.1840554
[3,] -0.40155454 -0.6503368 -0.53979845 -0.3193533
[4,]   0.12576714 -0.8208262  0.49453453 -0.3408145

sqrt(eigen(R1)$values)
[1] 0.4094346 0.2608864
```

for the personal variables. It appears that relatively older and medicated people tend to have a lower depression score, but perceive their health as relatively poor, while relatively younger but educated people have the opposite health perception. (I am grateful to Krzanowski, 2004, for this interpretation.)

The second canonical correlation is 0.261 which is again significant (see Exercise 8.3 and 8.4). The corresponding canonical variates are

$$u_2 = 0.95 \text{ CESD} - 0.32 \text{ Health},$$
$$v_2 = 0.620 \text{ Gender} - 0.630 \text{ Age} - 0.650 \text{ Education} - 0.821 \text{ Income}.$$

Since the higher value of the gender variable is for females, the interpretation here is that relatively young, poor, and uneducated females are associated with higher depression scores and, to a lesser extent, with poor perceived health (again this interpretation is due to Krzanowski, 2004).

8.4 Summary

Canonical correlation analysis has the reputation of being the most difficult multivariate technique to interpret. In many respects it is a well earned reputation! Certainly one has to know the variables involved very well to have any hope of extracting a convincing explanation. But in some circumstances (the heads measurement data is an example), CCA does provide a useful description of the association between two sets of variables.

Exercises

8.1 If $\mathbf{x}$ is a q_1-dimensional vector and $\mathbf{y}$ a q_2-dimensional vector, show that they linear combinations $\mathbf{a}'\mathbf{x}$ and $\mathbf{b}'\mathbf{y}$ have correlation

$$p(\mathbf{a}, \mathbf{b}) = \frac{\mathbf{a}'\Sigma_{12}\mathbf{b}}{(\mathbf{a}'\Sigma_{11}\mathbf{a}\mathbf{b}'\Sigma_{22}\mathbf{b})^{1/2}},$$

where Σ_{11} is the covariance matrix of the $\mathbf{x}$ variables, Σ_{22} the corresponding matrix for the $\mathbf{y}$ variables, and Σ_{12} the covariances across the two sets of variables.

8.2 Table 8.6 contains data from O'Sullivan and Mahon (1966) (data also given in Rencher, 1995), giving measurements on blood glucose for 52 women. The y's represent fasting glucose measurements on three occasions and the x's are glucose measurements one hour after sugar intake. Investigate the relationship between the two sets of variables using canonical correlation analysis.

8.3 Not all canonical correlations may be statistically significant. An approximate test proposed by Bartlett (1947) can be used to determine how many significant

Table 8.6 Blood Glucose Measurements on Three Occasions. From *Methods of Multivariate Analysis*, Rencher, A.C. Copyright © 1995. Reprinted with permission of John Wiley & Sons, Inc.

Fasting			One hour after sugar intake		
y_1	y_2	y_3	x_1	x_2	x_3
60	69	62	97	69	98
56	53	84	103	78	107
80	69	76	66	99	130
55	80	90	80	85	114
62	75	68	116	130	91
74	64	70	109	101	103
64	71	66	77	102	130
73	70	64	115	110	109
68	67	75	76	85	119
69	82	74	72	133	127
60	67	61	130	134	121
70	74	78	150	158	100
66	74	78	150	131	142
83	70	74	99	98	105
68	66	90	119	85	109
78	63	75	164	98	138
103	77	77	160	117	121
77	68	74	144	71	153
66	77	68	77	82	89
70	70	72	114	93	122
75	65	71	77	70	109
91	74	93	118	115	150
66	75	73	170	147	121
75	82	76	153	132	115
74	71	66	413	105	100
76	70	64	114	113	129
74	90	86	73	106	116
74	77	80	116	81	77
67	71	69	63	87	70
78	75	80	105	132	80
64	66	71	86	94	133
67	71	69	63	87	70

(Continued)

Table 8.6 (*Continued*)

	Fasting		One hour after sugar intake		
78	75	80	105	132	80
64	66	71	83	94	133
67	71	69	63	87	70
78	75	80	105	132	80
64	66	71	83	94	133
71	80	76	81	87	86
63	75	73	120	89	59
90	103	74	107	109	101
60	76	61	99	111	98
48	77	75	113	124	97
66	93	97	136	112	122
74	70	76	109	88	105
60	74	71	72	90	71
63	75	66	130	101	90
66	80	86	130	117	144
77	67	74	83	92	107
70	67	100	150	142	146
73	76	81	119	120	119
78	90	77	122	155	149
73	68	90	102	90	122
72	83	68	104	69	96
65	60	70	119	94	89
52	70	76	92	94	100

NOTE: Measurements are in mg/100 ml.

relationships exist. The test statistic for testing that at least one canonical correlation is significant is

$$\Phi_0^2 = -\left\{ n - \frac{1}{2}(q_1 + q_2 + 1) \right\} \sum_{i=1}^{s} \log(1 - \lambda_i)$$

where the λ_i are the eigenvalues of $\mathbf{E}_1$ and $\mathbf{E}_2$. Under the null hypothesis that all correlations are zero Φ_0^2 has a chi-square distribution with $q_1 \times q_2$ degrees of freedom. Write R and S-PLUS code to apply this test to the headsize data and to the depression data.

8.4   If the test in the previous exercise is significant, then the largest canonical correlation is removed and the residual is tested for significance using

$$\phi_1^2 = -\left\{n - \frac{1}{2}(q_1 + q_2+)\right\} \sum_{i=2}^{s} \log(1 - \lambda_i).$$

Under the hypothesis that all but the largest canonical correlation is zero ϕ_1^2 has a chi-square distribution with $(q_1 - 1)(q_2 - 1)$ degrees of freedom. Amend the function written for Exercise 8.3 to include this further test and then extend it to test for the significance of all the canonical correlations in both the headsize and depression data sets.

9
Analysis of Repeated Measures Data

9.1 Introduction

The multivariate data sets considered in previous chapters have involved measurements or observations on a number of different variables for each object or individual in the study. In this chapter, however, we will consider multivariate data of a different nature, namely that resulting from the repeated measurements of the same variable on each unit in the sample. Examples of such data are common in many disciplines. Often the repeated measurements arise from the passing of time (*longitudinal data*) but this is not always so. The two data sets in Tables 9.1 and 9.2 illustrate both possibilities. The first, taken from Crowder (1998), gives the loads required to produce slippage x of a timber specimen in a clamp. There are eight specimens, each with 15 repeated measurements. The second data set in Table 9.2 reported in Zerbe (1979) and also given in Davis (2002), consists of plasma inorganic phosphate measurements obtained from 13 control and 20 obese patients 0, 0.5, 1, 1.5, 2, and 3 hours after an oral glucose challenge.

The distinguishing feature of a repeated measures study is that the response variable of interest and a set of explanatory variables are measured several times on each individual in the study. The main objective in such a study is to characterize change in the repeated values of the response variable and to determine the explanatory variables most associated with any change. Because several observations of the response variable are made on the same individual, it is likely that the measurements will be correlated rather than independent, even after conditioning on the explanatory variables. Consequently repeated measures data require special methods of analysis, and models for such data need to include parameters linking the explanatory variables to the repeated measurements, parameters analogous to those in the usual multiple regression model (see Chapter 8), and, in addition parameters that account for the correlational structure of the repeated measurements. It is the former parameters that are generally of most interest, with the latter often being regarded as *nuisance parameters*. But providing an adequate model for the correlational structure of the repeated measures is necessary to avoid misleading inferences about the parameters that are of most importance to the researcher.

Table 9.1 Data Giving Loads Needed for a Given Slippage in 8 Specimens of Timber. From Nonlinear Growth Curves, Crowder, M.J., in *Encyclopedia of Biostatistics*, Armitage, P. and Colton, T. (Eds), Vol. 4, pp 3012–3014. Copyright © John Wiley & Sons Limited. Reproduced with permission.

Specimen	Slippage														
	0.0	0.10	0.20	0.30	0.40	0.50	0.60	0.70	0.80	0.90	1.00	1.20	1.40	1.60	1.80
1	0.0	2.38	4.34	6.64	8.05	9.78	10.97	12.05	12.98	13.94	14.74	16.13	17.98	19.52	19.97
2	0.0	2.69	4.75	7.04	9.20	10.94	12.23	13.19	14.08	14.66	15.37	16.89	17.78	18.41	18.97
3	0.0	2.85	4.89	6.61	8.09	9.72	11.03	12.14	13.18	14.12	15.09	16.68	17.94	18.22	19.40
4	0.0	2.46	4.28	5.88	7.43	8.32	9.92	11.10	12.23	13.24	14.19	16.07	17.43	18.36	18.93
5	0.0	2.97	4.68	6.66	8.11	9.64	11.06	12.25	13.35	14.54	15.53	17.38	18.76	19.81	20.62
6	0.0	3.96	6.46	8.14	9.35	10.72	11.84	12.85	13.83	14.85	15.79	17.39	18.44	19.46	20.05
7	0.0	3.17	5.33	7.14	8.29	9.86	11.07	12.13	13.15	14.09	15.11	16.69	17.69	18.71	19.54
8	0.0	3.36	5.45	7.08	8.32	9.91	11.06	12.21	13.16	14.05	14.96	16.24	17.34	18.23	18.87

Table 9.2 Plasma Inorganic Phosphate Levels from 33 Subjects. From *Statistical Methods for the Analysis of Repeated Measurements*, Davis, C.F., 2002. Copyright Springer-Verlag New York Inc. Reprinted with permission.

					Hours after glucose challenge				
Group control	ID	0	0.5	1	1.5	2	3	4	5
	1	4.3	3.3	3.0	2.6	2.2	2.5	3.4	4.4
	2	3.7	2.6	2.6	1.9	2.9	3.2	3.1	3.9
	3	4.0	4.1	3.1	2.3	2.9	3.1	3.9	4.0
	4	3.6	3.0	2.2	2.8	2.9	3.9	3.8	4.0
	5	4.1	3.8	2.1	3.0	3.6	3.4	3.6	3.7
	6	3.8	2.2	2.0	2.6	3.8	3.6	3.0	3.5
	7	3.8	3.0	2.4	2.5	3.1	3.4	3.5	3.7
	8	4.4	3.9	2.8	2.1	3.6	3.8	4.0	3.9
	9	5.0	4.0	3.4	3.4	3.3	3.6	4.0	4.3
	10	3.7	3.1	2.9	2.2	1.5	2.3	2.7	2.8
	11	3.7	2.6	2.6	2.3	2.9	2.2	3.1	3.9
	12	4.4	3.7	3.1	3.2	3.7	4.3	3.9	4.8
	13	4.7	3.1	3.2	3.3	3.2	4.2	3.7	4.3
	14	4.3	3.3	3.0	2.6	2.2	2.5	2.4	3.4
	15	5.0	4.9	4.1	3.7	3.7	4.1	4.7	4.9
	16	4.6	4.4	3.9	3.9	3.7	4.2	4.8	5.0
	17	4.3	3.9	3.1	3.1	3.1	3.1	3.6	4.0
	18	3.1	3.1	3.3	2.6	2.6	1.9	2.3	2.7
	19	4.8	5.0	2.9	2.8	2.2	3.1	3.5	3.6
	20	3.7	3.1	3.3	2.8	2.9	3.6	4.3	4.4
Obese	21	5.4	4.7	3.9	4.1	2.8	3.7	3.5	3.7
	22	3.0	2.5	2.3	2.2	2.1	2.6	3.2	3.5
	23	4.9	5.0	4.1	3.7	3.7	4.1	4.7	4.9
	24	4.8	4.3	4.7	4.6	4.7	3.7	3.6	3.9
	25	4.4	4.2	4.2	3.4	3.5	3.4	3.8	4.0
	26	4.9	4.3	4.0	4.0	3.3	4.1	4.2	4.3
	27	5.1	4.1	4.6	4.1	3.4	4.2	4.4	4.9
	28	4.8	4.6	4.6	4.4	4.1	4.0	3.8	3.8
	29	4.2	3.5	3.8	3.6	3.3	3.1	3.5	3.9
	30	6.6	6.1	5.2	4.1	4.3	3.8	4.2	4.8
	31	3.6	3.4	3.1	2.8	2.1	2.4	2.5	3.5
	32	4.5	4.0	3.7	3.3	2.4	2.3	3.1	3.3
	33	4.6	4.4	3.8	3.8	3.8	3.6	3.8	3.8

Over the last decade methodology for the analysis of repeated measures data has been the subject of much research and development, and there are now a variety of powerful techniques available. A comprehensive account of these methods is given in Diggle et al. (2002) and Davis (2002). Here we will concentrate on a single class of methods, *linear mixed effects models.*

9.2 Linear Mixed Effects Models for Repeated Measures Data

Linear mixed effects models for repeated measures data formalize the sensible idea that an individual's pattern of responses is likely to depend on many characteristics of that individual, including some that are unobserved. These unobserved variables are then included in the model as random variables, that is, random effects. The essential feature of the model is that correlation amongst the repeated measurements on the same unit arises from stored, unobserved variables. Conditional on the values of the random effects, the repeated measurements are assumed to be independent, the so-called *local independence* assumption.

Linear mixed effects models are introduced in Display 9.1 in the context of the timber slippage data in Table 9.1 by describing two commonly used models, the *random intercept* and *random intercept and slope* models.

Display 9.1
Two Simple Linear Mixed Effects Models

- Let y_{ij} represent the load in specimen i needed to produce a slippage of x_j, with $i = 1, \ldots, 8$ and $j = 1, \ldots, 15$. A possible model for the y_{ij} might be

$$y_{ij} = \beta_0 + \beta_1 x_j + u_i + \varepsilon_{ij} \qquad \text{(A)}$$

- Here the total residual that would be present in the usual linear regression model has been partitioned into a subject-specific random component u_i, which is constant over time plus a residual ε_{ij}, which varies randomly over time. The u_i are assumed to be normally distributed with zero mean and variance σ_u^2. Similarly the ε_{ij} are assumed to be normally distributed with zero mean and variance σ^2. The u_i and the ε_{ij} are assumed to be independent of each other and of the x_j.
- The model in (A) is known as a *random intercept model*, the u_i being the random intercepts. The repeated measurements for a specimen vary about that specimen's own regression line which can differ in intercept but not in slope from the regression lines of other specimens. The random effects model possible heterogeneity in the intercepts of the individuals.
- In this model slippage has a fixed effect.
- The random intercept model implies that the total variance of each repeated measurement is

$$\text{Var}(u_i + \varepsilon_{ij}) = \sigma_u^2 + \sigma^2.$$

- Due to this decomposition of the total residual variance into a between-subject component, σ_u^2, and a within-subject component, σ^2, the model is sometimes referred to as a *variance component model*.
- The covariance between the total residuals at two slippage levels j and j' in the same specimen i is

$$\text{Cov}(u_i + \varepsilon_{ij}, u_i + \varepsilon_{ij'}) = \sigma_u^2.$$

- Note that these covariances are induced by the shared random intercept; for specimens with $u_i > 0$, the total residuals will tend to be greater than the mean, for specimens with $u_i < 0$ they will tend to be less than the mean.
- It follows from the two relations above that the residual correlations are given by

$$\text{Cor}(u_i + \varepsilon_{ij}, u_i + \varepsilon_{ij'}) = \frac{\sigma_u^2}{\sigma_u^2 + \sigma^2}.$$

- This is an *intraclass correlation* interpreted as the proportion of the total residual variance that is due to residual variability between subjects.
- A random intercept model constrains the variance of each repeated measure to be the same and the covariance between any pair of measurements to be equal. This is usually called the *compound symmetry* structure.
- These constraints are often not realistic for repeated measures data. For example, for longitudinal data it is more common for measures taken closer to each other in time to be more highly correlated than those taken further apart. In addition the variances of the later repeated measures are often greater than those taken earlier.
- Consequently for many such data sets the random intercept model will not do justice to the observed pattern of covariances between the repeated measures. A model that allows a more realistic structure for the covariances is one that allows heterogeneity in both slopes and intercepts, the *random slope and intercept model*.
- In this model there are two types of random effects, the first modelling heterogeneity in intercepts, u_{i1}, and the second modelling heterogeneity in slopes, u_{i2}.
- Explicitly the model is

$$y_{ij} = \beta_0 + \beta_1 x_j + u_{i1} + u_{i2} x_j + \varepsilon_{ij}, \qquad (B)$$

where the parameters are not, of course, the same as in (A).
- The two random effects are assumed to have a bivariate normal distribution with zero means for both variables, variances $\sigma_{u_1}^2$, $\sigma_{u_2}^2$ and covariance $\sigma_{u_1 u_2}$.
- With this model the total residual is $u_{i1} + u_{i2} x_j + \varepsilon_{ij}$ with variance

$$\text{Var}(u_{i1} + u_{i2} x_j + \varepsilon_{ij}) = \sigma_{u_1}^2 + 2\sigma_{u_1 u_2} x_j + \sigma_{u_2}^2 x_j^2 + \sigma^2,$$

which is no longer constant for different values of x_j.

- Similarly the covariance between two total residuals of the same individual

$$\text{Cov}(u_{i1} + x_j u_{i2} + \varepsilon_{ij}, u_{i1} + u_{i2}x_{j'} + \varepsilon_{ij'})$$
$$= \sigma_{u_1}^2 + \sigma_{u_1 u_2}(x_j + x_{j'}) + \sigma_{u_2}^2 x_j x_{j'}$$

  is not constrained to be the same for all pairs j and j'.
- Linear mixed-effects models can be estimated by maximum likelihood. How-ever, this method tends to underestimate the variance components. A modified version of maximum likelihood, known as *restricted maximum likelihood*, is therefore often recommended; this provides consistent estimates of the vari-ance components. Details are given in Diggle et al. (2002) and Longford (1993).
- It should also be noted that re-estimating the models after adding or subtract-ing a constant from x_j (e.g., its mean), will lead to different variance and covariance estimates, but will not affect fixed effects.
- Competing linear mixed-effects models can be compared using a likelihood ratio test. If, however, the models have been estimated by restricted maximum likelihood this test can only be used if both models have the same set of fixed effects (see Longford, 1993).

Assuming that the data is available as shown in Table 9.1 as the matrix `timber`, we first need to rearrange it into what is known as the *long form* before we can apply the `lme` function that fits linear mixed effects models. This simply means that the repeated measurements are arranged "vertically" rather than horizontally as in Table 9.1. Suitable R and S-PLUS® code to make this rearrangement is

```
x<-c(0.0,0.1,0.2,0.3,0.4,0.5,0.6,0.7,0.8,0.9,1.0,1.2,1.4,
   1.6,1.8)
#
slippage<-rep(x,8)
loads<-as.vector(t(timber))
specimen<-rep(1:8,rep(15,8))
#
timber.dat<-data.frame(specimen,slippage,loads)
#
```

The rearranged data (`timber.dat`) are shown in Table 9.3. We can now fit the two models (A) and (B) as described in Display 9.1 and test one against the other using the `lme` function (in R the `nlme` library will first need to be loaded);

```
#in R use library(nlme)
attach(timber.dat)
#random intercept model
```

Table 9.3 Timber Data in "Long" Form

Observation	Specimen	Slippage	Load
1	1	0.0	0.00
2	1	0.1	2.38
3	1	0.2	4.34
4	1	0.3	6.64
5	1	0.4	8.05
6	1	0.5	9.78
7	1	0.6	10.97
8	1	0.7	12.05
9	1	0.8	12.98
10	1	0.9	13.94
11	1	1.0	14.74
12	1	1.2	16.13
13	1	1.4	17.98
14	1	1.6	19.52
15	1	1.8	19.97
16	2	0.0	0.00
17	2	0.1	2.69
18	2	0.2	4.75
19	2	0.3	7.04
20	2	0.4	9.20
21	2	0.5	10.94
22	2	0.6	12.23
23	2	0.7	13.19
24	2	0.8	14.08
25	2	0.9	14.66
26	2	1.0	15.37
27	2	1.2	16.89
28	2	1.4	17.78
29	2	1.6	18.41
30	2	1.8	18.97
31	3	0.0	0.00
32	3	0.1	2.85
33	3	0.2	4.89
34	3	0.3	6.61
35	3	0.4	8.09
36	3	0.5	9.72

(Continued)

Table 9.3 (*Continued*)

Observation	Specimen	Slippage	Load
37	3	0.6	11.03
38	3	0.7	12.14
39	3	0.8	13.18
40	3	0.9	14.12
41	3	1.0	15.09
42	3	1.2	16.68
43	3	1.4	17.94
44	3	1.6	18.22
45	3	1.8	19.40
46	4	0.0	0.00
47	4	0.1	2.46
48	4	0.2	4.28
49	4	0.3	5.88
50	4	0.4	7.43
51	4	0.5	8.32
52	4	0.6	9.92
53	4	0.7	11.10
54	4	0.8	12.23
55	4	0.9	13.24
56	4	1.0	14.19
57	4	1.2	16.07
58	4	1.4	17.43
59	4	1.6	18.36
60	4	1.8	18.93
61	5	0.0	0.00
62	5	0.1	2.97
63	5	0.2	4.68
64	5	0.3	6.66
65	5	0.4	8.11
66	5	0.5	9.64
67	5	0.6	11.06
68	5	0.7	12.25
69	5	0.8	13.35
70	5	0.9	14.54
71	5	1.0	15.53
72	5	1.2	17.38

(*Continued*)

Table 9.3 (*Continued*)

Observation	Specimen	Slippage	Load
73	5	1.4	18.76
74	5	1.6	19.81
75	5	1.8	20.62
76	6	0.0	0.00
77	6	0.1	3.96
78	6	0.2	6.46
79	6	0.3	8.14
80	6	0.4	9.35
81	6	0.5	10.72
82	6	0.6	11.84
83	6	0.7	12.85
84	6	0.8	13.83
85	6	0.9	14.85
86	6	1.0	15.79
87	6	1.2	17.39
88	6	1.4	18.44
89	6	1.6	19.46
90	6	1.8	20.05
91	7	0.0	0.00
92	7	0.1	3.17
93	7	0.2	5.33
94	7	0.3	7.14
95	7	0.4	8.29
96	7	0.5	9.86
97	7	0.6	11.07
98	7	0.7	12.13
99	7	0.8	13.15
100	7	0.9	14.09
101	7	1.0	15.11
102	7	1.2	16.69
103	7	1.4	17.69
104	7	1.6	18.71
105	7	1.8	19.54
106	8	0.0	0.00
107	8	0.1	3.36
108	8	0.2	5.45

(*Continued*)

Table 9.3 (*Continued*)

Observation	Specimen	Slippage	Load
109	8	0.3	7.08
110	8	0.4	8.32
111	8	0.5	9.91
112	8	0.6	11.06
113	8	0.7	12.21
114	8	0.8	13.16
115	8	0.9	14.05
116	8	1.0	14.96
117	8	1.2	16.24
118	8	1.4	17.34
119	8	1.6	18.23
120	8	1.8	18.87

```
timber.lme<-
   lme(loads~slippage,random=~1|specimen,data=timber.dat,
   method="ML")
#random intercept and slope model
timber.lme1<-
   lme(loads~slippage,random=~slippage|specimen,data
   =timber.dat, method="ML")
#compare two models
anova(timber.lme,timber.lme1)
```

The *p*-value associated with the likelihood ratio test is very small indicating that the random intercept and slope model is to be preferred over the simpler random intercept model for these data. The results from this model found from

```
summary(timber.lme1)
```

are shown in Table 9.4. The regression coefficient for slippage is highly significant. We can find the predicted values under this model and then plot them alongside a plot of the raw data using the following R and S-PLUS code:

```
predictions<-matrix(predict(timber.lme1),ncol=15,byrow=T)
par(mfrow=c(1,2))
matplot(x,t(timber),type="l",col=1,xlab="Slippage",
   ylab="Load",lty=1,
```

Table 9.4 Results of Random Intercept and Slope Model for the Timber Data

Effect	Estimated reg coeff	SE	DF	*t*-value	*p*-value
Intercept	3.52	0.26	111	13.30	<0.0001
Slippage	10.37	0.28	111	36.59	<0.0001

$\hat{\sigma}_{u_1} = 0.042, \hat{\sigma}_{u_2} = 0.014, \hat{\sigma} = 1.64.$

```
ylim=c(0,25))
title("(a)")
matplot(x,t(predictions),type="l",col=1,xlab="Slippage",
    ylab="Load",lty=1,
ylim=c(0,25))
title("(b)")
```

The resulting plot is shown in Figure 9.1 Clearly the fit is not good. In fact, under the random intercept and slope model the predicted values for each specimen are almost identical, reflecting the fact that the estimated variances of both random effects are essentially zero.

The plot of the observed values in Figure 9.1 shows that a quadratic term in slippage is essential in any model for these data. Including this as a fixed effect, the required model is

$$y_{ij} = \beta_0 + \beta_1 x_j + \beta_2 x_j^2 + u_{i1} + u_{i2} x_j + \varepsilon_{ij}. \qquad (9.1)$$

The necessary R and S-PLUS code to fit this model and test it against the previous random intercept and slope model is

```
timber.lme2<-lme(loads~slippage+I(slippage*slippage),
random=~slippage|specimen,data=timber.dat,method="ML")
anova(timber.lme1,timber.lme2)
```

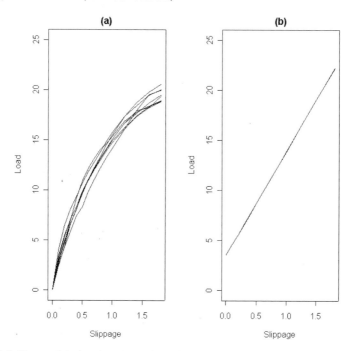

Figure 9.1 Observed timber data (a) and predicted values from random intercept and slope model (b).

Table 9.5 Results of Random Intercept and Slope Model with a Fixed Quadratic Effect for Slippage for Timber Data

Effect	Estimated reg coeff	Sd S.E	DF	t-value	p-value
Intercept	0.94	0.21	110	4.52	<0.0001
Slippage	19.89	0.33	110	61.11	<0.0001
Slippage^2	−5.43	0.17	110	−32.62	<0.0001

$\hat{\sigma}_{u_1} = 0.049, \hat{\sigma}_{u_2} = 0.032, \hat{\sigma} = 0.50.$

The p-value from the likelihood ratio test is less than 0.0001 indicating that the model that includes a quadratic term does provide a much improved fit. The results from this model are shown in Table 9.5. Both the linear and quadratic effects of slippage are highly significant.

We can now produce a similar plot to that in Figure 9.1 but showing the predicted values from the model in (9.1). The code is similar to that given above and so is not repeated again here. The resulting plot is shown in Figure 9.2. Clearly the model describes the data more satisfactorily although there remains an obvious problem which is taken up in Exercise 9.1.

Now we can move on to consider the data in Table 9.2, which we assume are available as the matrix `plasma`. Here we will begin by plotting the data so that we

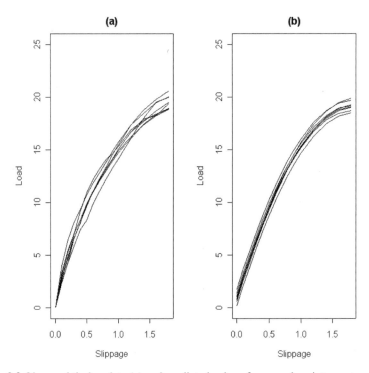

Figure 9.2 Observed timber data (a) and predicted values from random intercept and slope model that includes a quadratic effect for slippage (b).

get some ideas as to what form of linear mixed effect model might be appropriate. First we plot the raw data separately for the control and the obese groups using the following code:

```
par(mfrow=c(1,2))
matplot(matrix(c(0.0,0.5,1.0,1.5,2.0,2.5,3.0,4.0),ncol=1),
t(plasma[1:13,]),type="l",col=1,lty=1,
xlab="Time (hours after oral glucose challenge)",
    ylab="Plasma inorganic phosphate",ylim=c(1,7))
title("Control")
matplot(matrix(c(0.0,0.5,1.0,1.5,2.0,2.5,3.0,4.0),ncol=1),
t(plasma[14:33,]),type="l",col=1,lty=1,
xlab="Time (hours after glucose challenge)",ylab="Plasma
    inorganic phosphate",ylim=c(1,7))
title("Obese")
```

This gives Figure 9.3. The profiles in both groups show some curvature, suggesting that a quadratic effect of time may be needed in any model. There also appears to be some difference in the shape of the curves in the two groups, suggesting perhaps the need to consider a group × time interaction.

Next we plot the scatterplot matrices of the repeated measurements for the two groups using;

```
pairs(plasma[1:13,])
pairs(plasma[14:33,])
```

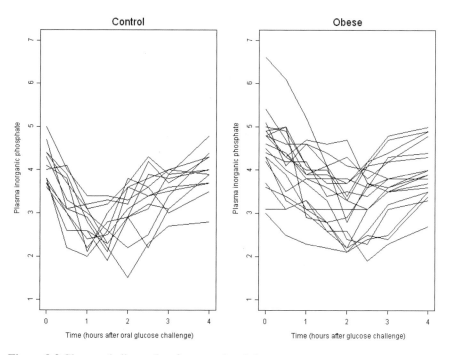

Figure 9.3 Glucose challenge data for control and obese groups.

The results are shown in Figures 9.4 and 9.5. Both plots indicate that the correlations of pairs of measurements made at different times differ so that the compound symmetry structure for these correlations is unlikely to be appropriate.

On the basis of the plots in Figure 9.3–9.5 we will begin by fitting the model in (9.1) with the addition, in this case, of an extra covariate, namely a dummy variable coding the group, control or obese, to which a subject belongs. We first need to put the data into the long form and combine with the appropriate group coding, subject number, and time. The necessary R and S-PLUS code for this is:

```
#
group<-rep(c(0,1),c(104,160))
#
time<-c(0.0,0.5,1.0,1.5,2.0,3.0,4.0,5.0)
time<-rep(time,33)
#
subject<-rep(1:33,rep(8,33))
plasma.dat<-cbind(subject,time,group,as.vector(t(plasma)))
dimnames(plasma.dat)<-list(NULL,c("Subject","Time","Group",
    "Plasma"))
plasma.df<-as.data.frame(plasma.dat)
plasma.df$Group<-factor(plasma.df$Group,levels=c(0,1),
    labels=c("Control","Obese"))
attach(plasma.df)
```

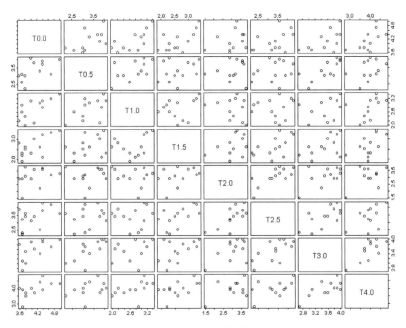

Figure 9.4 Scatterplot matrix for control group in Table 9.2.

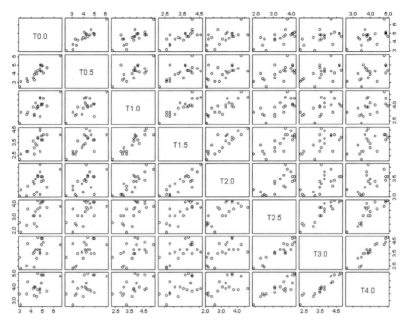

Figure 9.5 Scatterplot matrix for obese group in Table 9.2.

The first part of the rearranged data is shown in Table 9.6. We can fit the required model using

```
plasma.lme1<-lme(Plasma~Time+I(Time*Time)+Group,random
    =~Time|Subject,
data=plasma.df,method="ML")
summary(plasma.lme1)
```

The results are shown in Table 9.7. The regression coefficients for linear and quadratic time are both highly significant. The group effect just fails to reach significance at the 5% level. A confidence interval for the group effect is obtained from $0.38 \pm 2.04 \times 0.19$ giving $[-0.001, 0.767]$. (In S-PLUS the group effect and its standard error will be half those given in R corresponding to the group levels being coded by default as -1 and 1. This can be changed by use of the contr.treatment function.)

Here to demonstrate what happens if we make a very misleading assumption about the correlational structure of the repeated measurements, we will compare the results in Table 9.7 with those obtained if we assume that the repeated measurements are independent. The independence model can be fitted in the usual way with the lm function (see Chapter 8):

```
summary(lm(Plasma~Time+I(Time*Time)+Group,data=plasma.df))
```

The results are shown in Table 9.8. We see that under the independence assumption the standard error for the group effect is about one-half of that given in Table 9.7 and

Table 9.6 Part of Glucose Challenge Data in "Long" Form

	Subject	Time	Group	Plasma
1	1	0.0	Control	4.3
2	1	0.5	Control	3.3
3	1	1.0	Control	3.0
4	1	1.5	Control	2.6
5	1	2.0	Control	2.2
6	1	3.0	Control	2.5
7	1	4.0	Control	3.4
8	1	5.0	Control	4.4
9	2	0.0	Control	3.7
10	2	0.5	Control	2.6
11	2	1.0	Control	2.6
12	2	1.5	Control	1.9
13	2	2.0	Control	2.9
14	2	3.0	Control	3.2
15	2	4.0	Control	3.1
16	2	5.0	Control	3.9
17	3	0.0	Control	4.0
18	3	0.5	Control	4.1
19	3	1.0	Control	3.1
20	3	1.5	Control	2.3

if used would lead to the claim of strong evidence of a difference between control and obese patients.

We will now plot the predicted values from the fitted linear mixed effects model for each group using

```
predictions<-matrix(predict(plasma.lme1),ncol=8,byrow=T)
par(mfrow=c(1,2))
matplot(matrix(c(0.0,0.5,1,1.5,2,3,4,5),ncol=1),
t(predictions[1:13,]),type="l",lty=1,col=1,
xlab="Time (hours after glucose challenge)",ylab="Plasma
   inorganic phosphate",ylim=c(1,7))
title("Control")
matplot(matrix(c(0.0,0.5,1,1.5,2,3,4,5),ncol=1),
```

Table 9.7 Results from Random Slope and Intercept Model with Fixed Quadratic Time Effect Fitted to Glucose Challenge Data

Effect	Estimated reg coeff	SE	DF	t-value	p-value
Intercept	3.95	0.17	229	23.74	<0.0001
Time	−0.83	0.06	229	−13.34	<0.0001
Time^2	0.16	0.01	229	14.47	<0.0001
Group	0.38	0.19	31	2.03	0.051

$\hat{\sigma}_{u_1} = 0.61, \hat{\sigma}_{u_2} = 0.12, \hat{\sigma} = 0.42.$

Table 9.8 Results from Independence Model Fitted to Glucose Challenge Data

Effect	Estimated reg coeff	Sd S.E	t-value	p-value
Intercept	3.91	0.11	36.25	<0.0001
Time	−0.83	0.10	−8.65	<0.0001
Time^2	0.16	0.02	8.80	<0.0001
Group	0.46	0.09	5.24	<0.0001

```
t(predictions[14:33,]),type="l",lty=1,col=1,
xlab="Time (hours after glucose challenge)",ylab="Plasma
    inorganic phosphate",ylim=c(1,7))
title("Obese")
```

This gives Figure 9.6. We can see that the model has captured the profiles of the control group relatively well but not those of the obese group. We need to consider a further model that contains a group × time interaction.

The required model can be fitted and tested against the previous model using

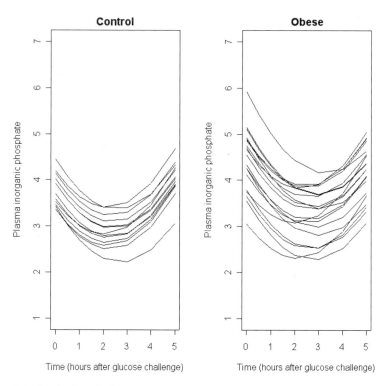

Figure 9.6 Fitted values from random intercept and slope model with fixed quadratic effect for glucose challenge data.

```
plasma.lme2<-lme(Plasma~Time*Group+I(Time*Time),random
    =~Time|Subject,
data=plasma.df,method="ML")
#
anova(plasma.lme1,plasma.lme2)
```

The *p*-value associated with the likelihood ratio test is 0.0011, indicating that the model containing the interaction term is to be preferred. The results for this model are given in Table 9.9. The interaction effect is highly significant. The fitted values from this model are shown in Figure 9.7 (the code is very similar to that given for producing Figure 9.6). The plot shows that the new model has produced predicted values that more accurately reflect the raw data plotted in Figure 9.3. The predicted profiles for the obese group are "flatter" as required.

We can check the assumptions of the final model fitted to the glucose challenge data, that is, the normality of the random effect terms and the residuals by first using the random.effects function to *predict* the former and the resid function to calculate the differences between the observed data values and the fitted values, and then using normal probability plots on each. How the random effects are predicted is explained briefly in Display 9.2. The necessary R and S-PLUS code to obtain the effects, residuals and plots is as follows:

```
res.int<-random.effects(plasma.lme2)[,1]
res.int
res.slope<-random.effects(plasma.lme2)[,2]
par(mfrow=c(1,3))
qqnorm(res.int,ylab="Estimated random intercepts",
    main="Random intercepts")
qqnorm(res.slope,ylab="Estimated random slopes",
    main="Random slopes")
resids<-resid(plasma.lme2)
qqnorm(resids,ylab="Estimated residuals",main="Residuals")
```

The resulting plot is shown in Figure 9.8. The plot of the residuals is linear as required, but there is some slight deviation from linearity for each of the predicted random effects.

Table 9.9 Results from Random Intercept Slope and Model with Quadratic Time Effect and Group × Time Interaction Fitted to Glucose Challenge Data

Effect	Estimated reg coeff	SE	DF	*t*-value	*p*-value
Intercept	3.70	0.18	228	20.71	<0.0001
Time	−0.73	0.07	228	−10.90	<0.0001
Time^2	0.16	0.01	228	14.44	<0.0001
Group	0.81	0.22	31	3.60	0.0011
Group × time	−0.16	0.05	228	−3.51	0.0005

$\hat{\sigma}_{u_1} = 0.57, \hat{\sigma}_{u_2} = 0.09, \hat{\sigma} = 0.42.$

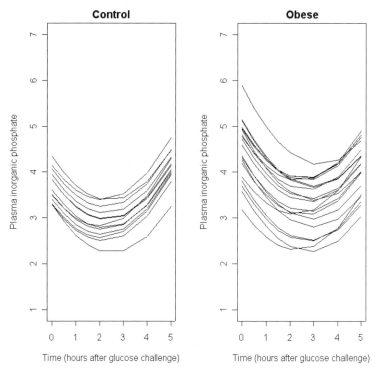

Figure 9.7 Fitted values from random intercept and slope model with fixed quadratic effect and group × time interaction for glucose challenge data.

Display 9.2
Prediction of Random Effects

- The random effects are not estimated as part of the model. However, having estimated the model, we can *predict* the values of the random effects.
- According to Bayes theorem, the *posterior probability* of the random effects is given by

$$\mathrm{Pr}(\mathbf{u}|\mathbf{y}, \mathbf{x}) = f(\mathbf{y}|\mathbf{u}, \mathbf{x})g(\mathbf{u}),$$

where $f(\mathbf{y}|\mathbf{u}, \mathbf{x})$ is the conditional density of the responses given the random effects and covariates (a product of normal densities) and $g(\mathbf{u})$ is the *prior* density of the random effects (multivariate normal). The means of this posterior distribution can be used as estimates of the random effects and are known as *empirical Bayes* estimates.

- The empirical Bayes estimator is also known as a shrinkage estimator because the predicted random effects are smaller in absolute value than their fixed-effect counterparts.
- *Best linear unbiased predictions* (BLUPs) are linear combinations of the responses that are unbiased estimators of the random effects and minimize the mean square error.

9.3 Dropouts in Longitudinal Data

A problem that frequently occurs when collecting longitudinal data is that some of the intended measurements are, for one reason or another, not made. In clinical trials, for example, some patients may miss one or more protocol scheduled visits after treatment has begun and so fail to have the required outcome measure taken. There will be other patients who do not complete the intended follow-up for some reason and drop out of the study before the end date specified in the protocol. Both situations result in missing values of the outcome measure. In the first case these are intermittent, but dropping out of the study implies that once an observation at a particular time point is missing so are all the remaining planned observations.

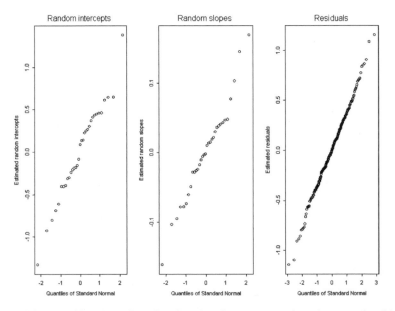

Figure 9.8 Probability plots of predicted random intercepts, random slopes, and residuals for final model fitted to glucose challenge data.

Many studies will contain missing values of both types, although in practice it is dropouts that cause most problems when turning to analyzing the resulting data set.

An example of a set of longitudinal data in which a number of patients have dropped out is given in Table 9.10. These data are essentially a subset of those collected in a clinical trial that is described in detail in Proudfoot et al. (2003). The trial was designed to assess the effectiveness of an interactive program using multimedia techniques for the delivery of cognitive behavioral therapy for depressed patients and known as Beating the Blues (BtB). In a randomized controlled trial of the program, patients with depression recruited in primary care were randomized to either the BtB program, or to Treatment as Usual (TAU). The outcome measure used in the trial was the Beck Depression Inventory II (Beck et al., 1996) with higher values indicating more depression. Measurements of this variable were made on five occasions, one prior to the start of treatment and at two monthly intervals after treatment began. In addition whether or not a participant in the trial was already taking antidepressant medication was noted along with the length of time they had been depressed.

To begin we shall graph the data here by plotting the boxplots of each of the five repeated measures separately for each treatment group. Assuming the data are available as the data frame btb.data the necessary code is

```
par(mfrow=c(2,1))
boxplot(btb.data[Treatment=="TAU",4],btb.data
   [Treatment=="TAU",5],btb.data[Treatment=="TAU",6],
btb.data[Treatment=="TAU",7],btb.data[Treatment=="TAU",8],
   names=c("BDIpre","BDI2m","BDI4m","BDI6m",
"BDI8m"),ylab="BDI",xlab="Visit",col=1)
title("TAU")
boxplot(btb.data[Treatment=="BtheB",4],btb.data
   [Treatment=="BtheB",5],btb.data[Treatment=="BtheB",6],
btb.data[Treatment=="BtheB",7],btb.data
   [Treatment=="BtheB",8],names=c("BDIpre","BDI2m","BDI4m",
"BDI6m","BDI8m"),ylab="BDI",xlab="Visit",col=1)
title("BtheB")
```

The resulting diagram is shown in Figure 9.9.

Figure 9.9 shows that there is decline in BDI values in both groups with perhaps the values in the BtheB group being lower at each postrandomization visit. We shall fit both random intercept and random intercept and slope models to the data including the pre-BDI values, treatment group, drugs, and length as fixed-effect covariates. First we need to rearrange the data into the long form using the following code:

```
n<-length(btb.data[,1])
#
BDI<-as.vector(t(btb.data[,c(5,6,7,8)]))
#
treat<-rep(btb.data[,3],rep(4,n))
subject<-rep(1:n,rep(4,n))
preBDI<-rep(btb.data[,4],rep(4,n))
drug<-rep(btb.data[,1],rep(4,n))
```

Table 9.10 Subset of Data from the Original BtB Trial

Sub	DRUG	Duration	Treatment	BDIpre	BDI2m	BDI3m	BDI5m	BDI8m
1	n	>6 m	TAU	29	2	2	NA	NA
2	y	>6 m	BtheB	32	16	24	17	20
3	y	<6 m	TAU	25	20	NA	NA	NA
4	n	>6 m	BtheB	21	17	16	10	9
5	y	>6 m	BtheB	26	23	NA	NA	NA
6	y	<6 m	BtheB	7	0	0	0	0
7	y	<6 m	TAU	17	7	7	3	7
8	n	>6 m	TAU	20	20	21	19	13
9	y	<6 m	BtheB	18	13	14	20	11
10	y	>6 m	BtheB	20	5	5	8	12
11	n	>6 m	TAU	30	32	24	12	2
12	y	<6 m	BtheB	49	35	NA	NA	NA
13	n	>6 m	TAU	26	27	23	NA	NA
14	y	>6 m	TAU	30	26	36	27	22
15	y	>6 m	BtheB	23	13	13	12	23
16	n	<6 m	TAU	16	13	3	2	0
17	n	>6 m	BtheB	30	30	29	NA	NA
18	n	<6 m	BtheB	13	8	8	7	6
19	n	>6 m	TAU	37	30	33	31	22
20	y	<6 m	BtheB	35	12	10	8	10
21	n	>6 m	BtheB	21	6	NA	NA	NA
22	n	<6 m	TAU	26	17	17	20	12
23	n	>6 m	TAU	29	22	10	NA	NA
24	n	>6 m	TAU	20	21	NA	NA	NA
25	n	>6 m	TAU	33	23	NA	NA	NA
26	n	>6 m	BtheB	19	12	13	NA	NA
27	y	<6 m	TAU	12	15	NA	NA	NA
28	y	>6 m	TAU	47	36	49	34	NA
29	y	>6 m	BtheB	36	6	0	0	2
30	n	<6 m	BtheB	10	8	6	3	3
31	n	<6 m	TAU	27	7	15	16	0
32	n	<6 m	BtheB	18	10	10	6	8
33	y	<6 m	BtheB	11	8	3	2	15
34	y	<6 m	BtheB	6	7	NA	NA	NA
35	y	>6 m	BtheB	44	24	20	29	14
36	n	<6 m	TAU	38	38	NA	NA	NA

(Continued)

Table 9.10 (*Continued*)

Sub	DRUG	Duration	Treatment	BDIpre	BDI2m	BDI3m	BDI5m	BDI8m
37	n	<6 m	TAU	21	14	20	1	8
38	y	>6 m	TAU	34	17	8	9	13
39	y	<6 m	BtheB	9	7	1	NA	NA
40	y	>6 m	TAU	38	27	19	20	30
41	y	<6 m	BtheB	46	40	NA	NA	NA
42	n	<6 m	TAU	20	19	18	19	18
43	y	>6 m	TAU	17	29	2	0	0
44	n	>6 m	BtheB	18	20	NA	NA	NA
45	y	>6 m	BtheB	42	1	8	10	6
46	n	<6 m	BtheB	30	30	NA	NA	NA
47	y	<6 m	BtheB	33	27	16	30	15
48	n	<6 m	BtheB	12	1	0	0	NA
49	y	<6 m	BtheB	2	5	NA	NA	NA
50	n	>6 m	TAU	36	42	49	47	40
51	n	<6 m	TAU	35	30	NA	NA	NA
52	n	<6 m	BtheB	23	20	NA	NA	NA
53	n	>6 m	TAU	31	48	38	38	37
54	y	<6 m	BtheB	8	5	7	NA	NA
55	y	<6 m	TAU	23	21	26	NA	NA
56	y	<6 m	BtheB	7	7	5	4	0
57	n	<6 m	TAU	14	13	14	NA	NA
58	n	<6 m	TAU	40	36	33	NA	NA
59	y	<6 m	BtheB	23	30	NA	NA	NA
60	n	>6 m	BtheB	14	3	NA	NA	NA
61	n	>6 m	TAU	22	20	16	24	16
62	n	>6 m	TAU	23	23	15	25	17
63	n	<6 m	TAU	15	7	13	13	NA
64	n	>6 m	TAU	8	12	11	26	NA
65	n	>6 m	BtheB	12	18	NA	NA	NA
66	n	>6 m	TAU	7	6	2	1	NA
67	y	<6 m	TAU	17	9	3	1	0
68	y	<6 m	BtheB	33	18	16	NA	NA
69	n	<6 m	TAU	27	20	NA	NA	NA
70	n	<6 m	BtheB	27	30	NA	NA	NA
71	n	<6 m	BtheB	9	6	10	1	0
72	n	>6 m	BtheB	40	30	12	NA	NA

(*Continued*)

Table 9.10 (*Continued*)

Sub	DRUG	Duration	Treatment	BDIpre	BDI2m	BDI3m	BDI5m	BDI8m
73	n	>6 m	TAU	11	8	7	NA	NA
74	n	<6 m	TAU	9	8	NA	NA	NA
75	n	>6 m	TAU	14	22	21	24	19
76	y	>6 m	BtheB	28	9	20	18	13
77	n	>6 m	BtheB	15	9	13	14	10
78	y	>6 m	BtheB	22	10	5	5	12
79	n	<6 m	TAU	23	9	NA	NA	NA
80	n	>6 m	TAU	21	22	24	23	22
81	n	>6 m	TAU	27	31	28	22	14
82	y	>6 m	BtheB	14	15	NA	NA	NA
83	n	>6 m	TAU	10	13	12	8	20
84	y	<6 m	TAU	21	9	6	7	1
85	y	>6 m	BtheB	46	36	53	NA	NA
86	n	>6 m	BtheB	36	14	7	15	15
87	y	>6 m	BtheB	23	17	NA	NA	NA
88	y	>6 m	TAU	35	0	6	0	1
89	y	<6 m	BtheB	33	13	13	10	8
90	n	<6 m	BtheB	19	4	27	1	2
91	n	<6 m	TAU	16	NA	NA	NA	NA
92	y	<6 m	BtheB	30	26	28	NA	NA
93	y	<6 m	BtheB	17	8	7	12	NA
94	n	>6 m	BtheB	19	4	3	3	3
95	n	>6 m	BtheB	16	11	4	2	3
96	y	>6 m	BtheB	16	16	10	10	8
97	y	<6 m	TAU	28	NA	NA	NA	NA
98	n	>6 m	BtheB	11	22	9	11	11
99	n	<6 m	TAU	13	5	5	0	6
100	y	<6 m	TAU	43	NA	NA	NA	NA

```
length<-rep(btb.data[,2],rep(4,n))
time<-rep(c(2,4,6,8),n)
#
#
btb.bdi<-data.frame(subject,treat,drug,length,preBDI,
   time,BDI)
#
attach(btb.bdi)
```

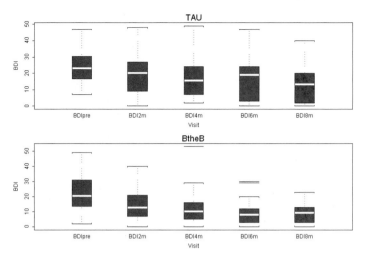

Figure 9.9 Boxplots for the repeated measures by treatment group for the BtheB data.

The resulting data frame `btb.bdi` contains a number of missing values and in applying the `lme` function these will need to be dropped. But notice it is only the missing values that are removed, *not* participants that have at least one missing value. All the available data is used in the model fitting process. We can fit the two models and test which is most appropriate using

```
btbbdi.fit1 <- lme(BDI ~ preBDI + time + treat + drug
    + length, method = "ML", random
= ~ 1 | subject, data= btb.bdi, na.action = na.omit)
btbbdi.fit2 <- lme(BDI ~ preBDI + time + treat + drug
    + length, method = "ML", random
= ~ time | subject,data = btb.bdi, na.action = na.omit)
anova(btbbdi.fit1, btbbdi.fit2)
```

This results in

Model		df	AIC	BIC	logLik	Test	L.Ratio	p-value
btbbdi.fit1	1	8	1886.624	1915.702	−935.3121			
btbbdi.fit2	2	10	1889.808	1926.156	−934.9040	1 vs 2	0.8160734	0.665

Clearly the simpler random intercept model is adequate for these data. The results from this model can be found using

```
summary(btbbdi.fit1)
```

and they are given in Table 9.11. Only time and the pre-BDI regression coefficients are significantly different from zero. In particular there is no occurring evidence of a treatment effect.

Table 9.11 Results from Random Intercept Model
Fitted to BtheB Data

Effect	Estimated reg coeff	SE	DF	t-value	p-value
Intercept	5.94	2.10	182	2.27	0.0986
Pre BDI	0.64	0.08	92	8.14	<.0001
Time	−0.72	0.15	182	−4.86	<.0001
Treatment	−2.37	1.68	92	−1.41	0.1616
Drug	−2.80	1.74	92	−1.61	0.1110
Duration	0.26	1.65	92	0.16	0.8769

$\hat{\sigma}_{u_1} = 6.95, \hat{\sigma} = 5.01.$

We now need to consider briefly how the dropouts may affect the analyses reported above. To understand the problems that patients dropping out can cause for the analysis of data from a longitudinal trial we need to consider a classification of dropout mechanisms first introduced by Rubin (1976). The type of mechanism involved has implications for which approaches to analysis are suitable and which are not. Rubin's suggested classification involves three types of dropout mechanism:

- *Dropout completely at random* (DCAR): Here the probability that a patient drops out does not depend on either the observed or missing values of the response. Consequently the observed (nonmissing) values effectively constitute a simple random sample of the values for all subjects. Possible examples include missing laboratory measurements because of a dropped testtube (if it was not dropped because of the knowledge of any measurement), the accidental death of a participant in a study, or a participant moving to another area. Intermittent missing values in a longitudinal data set, whereby a patient misses a clinic visit for transitory reasons ("went shopping instead" or the like) can reasonably be assumed to be DCAR. Completely random dropout causes the least problem for data analysis, but it is a strong assumption.
- *Dropout at random* (DAR): The dropout-at-random mechanism occurs when the probability of dropping out depends on the outcome measures that have been observed in the past, but given this information is conditionally independent of all the future (unrecorded) values of the outcome variable following dropout. Here "missingness" depends only on the observed data with the distribution of future values for a subject who drops out at a particular time being the same as the distribution of the future values of a subject who remains in at that time, if they have the same covariates and the same past history of outcome up to and including the specific time point. Murray and Findlay (1988) provide an example of this type of missing value from a study of hypertensive drugs in which the outcome measure was diastolic blood pressure. The protocol of the study specified that the participant was to be removed from the study when his/her blood pressure got too large. Here blood pressure at the

time of dropout was observed before the participant dropped out, so although the dropout mechanism is not DCAR since it depends on the values of blood pressure, it *is* DAR, because dropout depends only on the observed part of the data. A further example of a DAR mechanism is provided by Heitjan (1997), and involves a study in which the response measure is body mass index (BMI). Suppose that the measure is missing because subjects who had high body mass index values at earlier visits avoided being measured at later visits out of embarrassment, regardless of whether they had gained or lost weight in the intervening period. The missing values here are DAR but *not* DCAR; consequently methods applied to the data that assumed the latter might give misleading results (see later discussion).

• *Nonignorable* (sometimes referred to as *informative*): The final type of dropout mechanism is one where the probability of dropping out depends on the unrecorded missing values—observations are likely to be missing when the outcome values that would have been observed had the patient not dropped out, are systematically higher or lower than usual (corresponding perhaps to their condition becoming worse or improving). A nonmedical example is when individuals with lower income levels or very high incomes are less likely to provide their personal income in an interview. In a medical setting possible examples are a participant dropping out of a longitudinal study when his/her blood pressure became too high and this value was not observed, or when their pain become intolerable and we did not record the associated pain value. For the BDI example introduced above, if subjects were more likely to avoid being measured if they had put on extra weight since the last visit, then the data are nonignorably missing. Dealing with data containing missing values that result from this type of dropout mechanism is difficult. The correct analyses for such data must estimate the dependence of the missingness probability on the missing values. Models and software that attempt this are available (see, e.g., Diggle and Kenward, 1994) but their use is not routine and, in addition, it must be remembered that the associated parameter estimates can be unreliable.

Under what type of dropout mechanism are the mixed effects models considered in this chapter valid? The good news is that such models can be shown to give valid results under the relatively weak assumption that the dropout mechanism is DAR (see Carpenter et al., 2002). When the missing values are thought to be informative, any analysis is potentially problematical. But Diggle and Kenward (1994) have developed a modeling framework for longitudinal data with informative dropouts, in which random or completely random dropout mechanisms are also included as explicit models.

The essential feature of the procedure is a logistic regression model for the probability of dropping out, in which the explanatory variables can include previous values of the response variable, and, in addition, the *unobserved* value at dropout as a *latent* variable (i.e., an unobserved variable). In other words, the dropout probability is allowed to depend on both the *observed* measurement history and the unobserved value at dropout. This allows both a formal assessment of the type of dropout mechanism in the data, and the estimation of effects of interest, for example, treatment effects under different assumption about the dropout mechanism. A full account technical account of the model is given in Diggle and Kenward (1994) and a detailed example that uses the approach is described in Carpenter et al. (2002).

One of the problems for an investigator struggling to identify the dropout mechanism in a data set is that there are no routine methods to help, although a number of largely ad hoc graphical procedures can be used as described in Diggle (1998), Everitt (2002), and Carpenter (2002). Exercise 9.4 considers one of these.

9.4 Summary

Linear mixed effects models are extremely useful for modelling longitudinal data in particular and repeated measures data more generally. The models allow the correlations between the repeated measurements to be accounted for so that correct inferences can be drawn about the effects of covariates of interest on the repeated response values. In this chapter we have concentrated on responses that are continuous and conditional on the explanatory variables and random effects have a normal distribution. But random effects models can also be applied to nonnormal responses, for example, binary variables; see, for example, Everitt (2002).

The lack of independence of repeated measures data is what makes the modelling of such data a challenge. But even when only a single measurement of a response is involved, correlation can, in some circumstances, occur between the response values of different individuals and cause similar problems. As an example consider a randomized clinical trial in which subjects are recruited at multiple study centers. The multicenter design can help to provide adequate sample sizes and enhance the generalizability of the results. However factors that vary by center, including patient characteristics and medical practice patterns, may exert a sufficiently powerful effect to make inferences that ignore the "clustering" seriously misleading. Consequently it may be necessary to incorporate random effects for centers into the analysis.

Exercises

9.1   The final model fitted to the timber data did not constrain the fitted curves to go through the origin although this is clearly necessary. Fit an amended model where this constraint is satisfied and plot the new predicted values.

9.2   Investigate a further model for the glucose challenge data that allow a random quadratic effect.

9.3   Fit an independence model to the BtheB data and compare the estimated treatment effect confidence interval with that from the random intercept model described in the text.

9.4   Investigate whether there is any evidence of an interaction between treatment and time for the Beat the Blues data.

9.5 One very simple procedure for assessing the dropout mechanism suggested in Carpenter et al. (2002) involves plotting the observations for each treatment group, at each time point, differentiating between two categories of patients; those who do and those who do not attend their next scheduled visit. Any clear difference between the distributions of values for these two categories indicates that dropout is not completely at random. Produce such a plot for the Beat the Blues data.

Appendix
An Aide Memoir for R and S-PLUS®

1. Elementary Commands

Elementary commands consist of either expressions or assignments. For example, typing the expression

```
> 42 + 8
```

in the **Commands** window and pressing **Return** will produce the following output:

```
[1]    50
```

In the remainder of this chapter, we will show the command (preceded by the prompt >) and the output as they would appear in the **Commands** window together like this:

```
> 42 + 8
[1]    50
```

Instead of just evaluating an expression, we can assign the value to a scalar using the syntax scalar <- expression

```
> x <- 42 + 8
```

Longer commands can be split over several lines by pressing Return before the command is complete. To indicate waiting for completion of a command, a "+" occurs instead of the > prompt. For illustration, we break the line in the assignment above:

```
> x<-
+ 48+8
```

2. Vectors

A commonly used type of R and S-PLUS® object is a *vector*. Vectors may be created in several ways of which the most common is via the concentrate command, c, which combines all values given as arguments to the function into a vector. For example,

```
>x<-c(1, 2, 3,4)
>x
[1] 1 2 3 4
```

Here, the first command creates a vector and the second command, *x*, a short-form for print(x), causes the contents of the vector to be printed. (Note that R and S-PLUS are case sensitive, and so, for example, x and X are different objects.)
The number of elements of a vector can be determined using the length() function:

```
>length(x)
[1] 4
```

The c function can also be used to combine *strings* which are denoted by enclosing them in "." For example,

```
>names <-c ("Brian", "Sophia", "Harry")
>names
[1] "Brian" "Sophia" "Harry"
```

The c() function also works with a mixture of numeric and string values, but in this case, all elements in the resulting vector will be converted to strings as in the following.

```
> mix <-c(names, 55, 33)
> mix
[1] "Brian" "Sophia"    "Harry" "55"    "33"
```

Vectors consisting of regular sequences of numbers can be created using the seq() function. The general syntax of this function is seq(lower, upper, increment). Some examples are given below:

```
>seq (1, 5, 1)
[1] 1 2 3 4 5
>seq (2, 20, 2)
[1] 2 4 6 8 10 12 14 16 18 20
>x <-c(seq(1, 5, 1), seq (4, 20, 4))
>x
[1] 1 2 3 4 5 4 8 12 16 20
```

When the increment argument is one it can be left out of the command. The same applies to the lower value. More information about the seq function and all other R and S-PLUS functions can be found using the help facilities, e.g.,

```
>help(seq)
```

shows the following information:

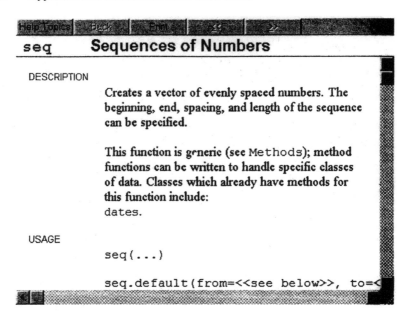

Sequences with increments of one can also be obtained using the syntax `first:last`, for example,

```
>1:5
[1] 1 2 3 4 5
```

A further useful function for creating vectors with regular patterns is the `rep` function, with general form `rep(pattern, number of times)`. For example,

```
>rep(10, 5)
[1] 10 10 10 10 10
>rep (1:3, 3)
[1] 1 2 3 1 2 3 1 2 3
> x <- rep(seq(5), 2)
> x
[1] 1 2 3 4 5 1 2 3 4 5
```

The second argument of `rep` can also be a vector of the same length as the first argument to indicate how often each element of the first argument is to be repeated as shown in the following;

```
> x <- rep(seq (3), c(1, 2, 3))
> x
[1] 1 2 2 3 3 3
```

Increasingly complex vectors can be built by repeated use of the `rep` function

```
> x <- rep (seq (3), rep (3, 3))
> x
[1] 1 1 1 2 2 2 3 3 3
```

We can access a particular element of a vector by giving the required position in square brackets- here are two examples

```
> x <- 1:5
> x[3]
[1] 3

>x[c(1, 4)]
[1] 1 4
```

A vector containing several elements of another vector can be obtained by giving a vector of required positions in square brackets:

```
> x[c(1, 3)]
[1] 1 3
> x[1:3]
[1] 1 2 3
```

We can carry out any of the arithmetic operations described in Table A.1 between two scalars, a vector and a scalar or two vectors. An arithmetic operation between two vectors returns a vector whose elements are the results of applying the operation to the corresponding elements of the original vectors. Some examples follow:

```
> x<- 1:3
> x+2
[1] 3 4 5

>x + x
[1] 2 4 6
> x* x
[1] 1 4 9
```

We can also apply mathematical functions such as the square root or logarithm, or the others listed in Table A.2, to vectors. The functions are simply applied to each element of the vector. For example,

```
> x <- 1:3
> sqrt (x*x)
[1] 1 2 3
```

Table A.1 Arithmetic Operators

Operator	Meaning	Expression	Result
+	Plus	$2 + 3$	5
−	Minus	$5 - 2$	3
*	Times	$5 * 2$	10
/	Divided by	$10/2$	5
∧	Power	$2 \wedge 3$	8

Table A.2 Common Functions

S-PLUS function	Meaning
sqrt ()	Square root
log ()	Natural logarithm
log10 ()	Logarithm base 10
exp ()	Exponential
abs ()	Absolute value
round ()	Round to nearest integer
ceiling ()	Round up
floor ()	Round down
sin (), cos (), tan ()	sine, cosine, tangent
asin (), acos (), atan ()	arc sine, arc cosine, arc tangent

3. Matrices

Matrix objects are frequently needed in R and S-PLUS and can be created by the use of the matrix function. The general syntax is

```
matrix (data, nrow, ncol, byrow = F)
```

The last argument specifies whether the matrix is to be filled row by row or column by column and takes on a logical value. The expression *byrow=F* indicates that F (false) is the default value. An example follows;

```
> x <-c(1, 2, 3)
> y <-c(4, 5, 6)
> xy <- matrix (c(x, y), nrow =2)
> xy
```

	[,1]	[,2]	[,3]
[1,]	1	3	5
[2,]	2	4	6

Here the number of columns is not specified and so is determined by simple division:

```
> xy <-matrix (c(x, y), nrow = 2, byrow =T)
xy
```

	[,1]	[,2]	[,3]
[1,]	1	2	3
[2,]	4	5	6

Here the matrix, is filled row-wise instead of by columns by setting the byrow argument to T for True. A square bracket with two numbers separated by a comma is used to refer to an element of a matrix. The first number specifies the row, and the second specifies the column.

```
>xy [1, 3]
[1] 3
```

The [i,] and [,j] nomenclature is used to refer to complete rows or columns of a matrix and can be used to extract particular rows or columns as shown in the following examples;

```
> xy [1,]
[1] 1 2 3
> xy [,2]
[1] 2 5

>xy [, c(1, 3)]

        [,1]     [,2]
[1,]     1        3
[2,]     4        6
```

As with vectors, arithmetic operations operate element by element when applied to matrices, for example;

```
> xy* xy

        [,1]     [,2]     [,3]
[1,]     1        4        9
[2,]     16       25       36
```

Matrix multiplication is performed using the %*% operation as here

```
> xy %*% t(xy)

        [,1]     [,2]
[1,]     14       32
[2,]     32       77
```

Here the matrix *xy* is multiplied by its transpose (obtained using the *t* () function). An attempt to apply matrix multiplication to *xy* by *xy* would, of course, result in an error message. It is usually extremely helpful to attach names to the rows and columns to a matrix. This can be done using the dimension() function. We shall illustrate this in Section 5 after we have covered list objects.

As with vectors, matrices can be formed from numeric and string objects, but in the resulting matrix, all elements will be strings as illustrated below:

```
> Mix <- matrix(c(names, 55, 32, 30), nrow = 2
+ byrow = T)
> Mix

        [,1]       [,2]        [,3]
[1,]    "Brian"    "Sophia"    "Harry"
[2,]    "55"       "32"        "30"
```

Higher dimensional matrices with up to eight dimensions can be defined using the array() function.

4. Logical Expressions

So far, we have mentioned values of type numeric or character (string). When a numeric value is missing, it is of type NA. (Complex numbers are also available.) Another type in R and S-PLUS is logical. There are two logical values, T (true) and

Table A.3 Logical Operators

Operator	Meaning
<	Less than
>	Greater than
<=	Less than or equal to
>=	Greater than or equal to
==	Equal to
!=	Not equal to
&	And
|	Or
!	not

F (false), and a number of logical operations that are extremely useful when making comparisons and choosing particular elements from vectors and matrices.

The symbols used for the logical operations are listed in Table A.3. We can use a logical expression to assign a logical value (T or F) to x:

```
> x <-3 = = 4
> x
[1]  F
> x <-3 < 4
> x
[1]  T
> x < - 3 = = 4 & 3 < 4
> x
[1]  F
> x < - 3 = = 4 | 3 < 4
> x
[1]  T
```

In addition to logical operators, there are also logical functions. Some examples are given below:

```
> is.numeric (3)
[1]  T
> is.character (3)
[1]  F
> is.character ("3")
[1]  T
> 1/0
[1]  Inf
>is.numeric (1/0)
[1]  T
>is.infinite (1/0)
[1]  T
```

Logical operators or functions operate on elements of vectors and matrices in the same say as arithmetic operators:

```
> is.na(c(1, 0, NA, 1))
[1]  F F T F
```

```
> ! is.na (c(1, 0, NA, 1))
[1] T T F T
> x <- seq(20)
> x <10
[1] T T T T T T T T T F F F F F F F F F F F
```

A logical vector can be used to extract a subset of elements from another vector as follows:

```
> x[x <10]
[1] 1 2 3 4 5 6 7 8 9
```

Here, the elements of the vector less than 10 are selected as the values corresponding to T in the vector $x < 10$. We can also select element in x depending on the values in another vector y:

```
> x <-seq(50)
> y <- c(rep(0, 10), rep(1, 40))
> x[y = =0]
[1]   1 2 3 4 5 6 7 8   9 10
```

5. List Objects

List objects allow any other R or S-PLUS objects to be linked together. For example,

```
>x<-seq(10)
>y<- matrix(seq(10), nrow = 5
>xylist<-list (x,y)
>xylist
[[1]]:
[1] 1 2 3 4 5 6 7 8 9 10

[[2]]:
        [,1]       [,2]
[1,]    1          6
[2,]    2          7
[3,]    3          8
[4,]    4          9
[5,]    5          10
```

Note the elements of the list are referred to by a double square brackets notation; so we can print the first component of the list using

```
>xylist[[1]]
[1] 1 2 3 4 5 6 7 8 9 10
```

The components of the list can also be given names and later referred using the list$name notation,

```
>xylist <-list (X=x, Y=y)
>xylist$X
[1] 1 2 3 4 5 6 7 8 9 10
```

List objects can, of course, include other list objects

```
>newlist<-list(xy=xylist, z=rep(0,10))
>newlist$xy
$X
[1] 1 2 3 4 5 6 7 8 9 10

$Y:
        [,1]      [,2]
[1,]    1     6
[2,]    2     7
[3,]    3     8
[4,]    4     9
[5,]    5     10

>newlist$z
[1] 0 0 0 0 0 0 0 0 0 0
```

The rows and columns of a matrix can be named using the dimnames() function and a list object

```
>x<-matrix(seq(12), nrow=4)
> dimnames(x)<-list(c("R1","R","R3","R4"),
+c("C1", "C2", "C3"))
>x

        C1      C2      C3
R1      1       5       9
R2      2       6       10
R3      3       7       11
R4      4       8       12
```

The names can be created more efficiently by using the paste() function, which combines different strings and numbers into a single string

```
> dimnames(x)<-list(paste("row", seq (4)),
+paste ("col", seq(3)))
>x
        col 1   col 2   col 3
row1    1       5       9
row2    2       6       10
row3    3       7       11
row4    4       8       12
```

Having named the rows and columns, we can, if required, refer to elements of the matrix using these names,

```
>x["row 1", "col 3"]
[1] 9
```

6. Data Frames

Data sets in R and S-PLUS are usually stored as matrices, which we have already met, or as data frames, which we shall describe here.

Data frames can bind vectors of different types together (e.g., numeric and character), retaining the correct type of each vector. In other respects, a data frame is like a matrix so that each vector should have the same number of elements. The syntax for creating a data frame is data.frame(vector1,vector 2, ...), and an example of how a small data frame can be created is as follows:

```
>height<-c(50, 70, 45, 80, 100)
>weight<-c(120, 140, 100, 200, 190)
>age<-c(20, 40, 41, 31, 33)
>names<-c("Bob", "Ted", "Alice", "Mary", "Sue")
sex<-c("Male", "Male", "Female", "Female")
>data<-data.frame(names, sex, height, weight, age)
>data
```

	names	sex	height	weight	age
1	Bob	Male	50	120	20
2	Ted	Male	70	140	40
3	Alice	Female	45	100	41
4	Mary	Female	80	200	31
5	Sue	Female	100	190	33

Particular parts of a data frame can be extracted in the same way as for matrices

```
>data[,c(1,2,5)]
```

	names	sex	age
1	Bob	Male	20
2	Ted	Male	40
3	Ali	Female	41
4	Mary	Female	31
5	Sue	Female	33

Column names can also be used

```
>data[,"age"]
[1] 20 40 41 31 33
```

Variables can also be accessed as in lists:

```
>data$age
[1] 20 40 41 31 33
```

It is, however, more convenient to "attach" a data frame and work with the column names directly, for example,

```
>attach (data)
>age
[1] 20 40 41 31 33
```

Note that the attach() command places the data frame in the 2nd position in the search path. If we assign a value to age, for example,

```
>age <-10
>age
[1] 10
```

This creates a new object in the first position of the search path that "masks" the age variable of the data frame. Variables can be removed from the first position in the search path using the rm() function:

```
>rm(age)
```

To change the value of age within the data frame, use the syntax

```
>data$age<-c(20, 30, 45, 32, 32)
```

References

Afifi, AA, Clark, VA and May, S. (2004) Computer-Aided Multivariate Analysis, (4th ed.). London: Chapman and Hall.

Alon, U, Barkai, N, Notterman, DA, Gish, K, Ybarra, S, Mack, D, and Levine, AJ (1999) Broad patterns of gene expressions revealed by clustering analysis of tumor and normal colon tissues probed by oligonucleotide arrays. Cell Biology, 99, 6745–6750.

Anderson, JA (1972) Separate sample logistic discrimination. Biometrika, 59, 19–35.

Banfield, JD and Raftery, AE (1993) Model-based Gaussian and non-Gaussian clustering. Biometrics, 49, 803–821.

Bartholomew, DJ (1987) Latent Variable Models and Factor Analysis. Oxford: Oxford University Press.

Bartlett, MS (1947) Multivariate analysis. Journal of the Royal Statistical Society B, 9, 176–197.

Becker, RA and Cleveland, WS (1994) S-PLUS Trellis Graphics User's Manual Version 3.3. Seattle: Mathsoft, Inc.

Benzécri, JP (1992) Correspondence Analysis Handbook. New York: Marcel Dekker.

Blackith, RE and Rayment, RA (1971) Multivariate Morphometrics. London: Academic Press.

Carpenter, J, Pocock, SJ, and Lamm, CJ (2002) Coping with missing data in clinical trials: A model-based approach applied to asthma trials. Statistics in Medicine, 21, 1043–1066.

Chambers, JM, Cleveland, WS, Kleiner, B, and Tukey, PA (1983) Graphical Methods for Data Analysis. Belmont, CA: Wadsworth.

Chatfield, C and Collins, AJ (1980) Introduction to Multivariate Analysis. London: Chapman and Hall.

Cleveland, WS (1979) Robust locally weighted regression and smoothing scatterplots. Journal of the American Statistical Association, 74, 829–836.

Cleveland, WS and McGill, ME (1987) Dynamic Graphics for Statistics. Belmont, CA: Wadsworth.

Crowder, MJ (1998) Nonlinear growth curve. In Encyclopedia of Biostatistics (eds. P Armitage and T Colton). Chichester: Wiley.

Dalgaard, P (2002) Introductory Statistics with R. New York: Springer.

Davis, CS (2002) Statistical Methods for the Analysis of Repeated Measurements. New York: Springer.

Dempster, AP, Laird, NM and Ruben, DB (1977) Maximum likelihood from incomplete data via the EM algorithm. Journal of the Royal Statistical Society, B, 39, 1–38.

de Leeuw, J (1985) Book review. Psychometrika, 50, 371–375.

Diggle, PJ (1998) Dealing with missing values in longitudinal studies. In Statistical Analysis of Medical Data (eds. BS Everitt and G Dunn). London: Arnold.

Diggle, PJ, Heagerty, P, Liang, KY, and Zeger, SL (2002) Analysis of Longitudinal Data. Oxford: Oxford University Press.

Diggle, PJ and Kenward, MG (1994) Informative dropout in longitudinal analysis (with discussion) Applied Statistics, 43, 49–93.

Dunn, G, Everitt, BS, and Pickles, A (1993) Modelling Covariances and Latent Variables Using EQS. London: Chapman and Hall.

Everitt, BS (1984) An Introduction to Latent Variable Models. Boca Raton, Florida: CRC/ Chapman and Hall.

Everitt, BS (1987) Statistics in Psychiatry. Statistical Science, 2, 107–134.

Everitt, BS (2002) Modern Medical Statistics: A Practical Guide. London: Arnold.

Everitt, BS and Bullmore, ET (1999) Mixture model mapping of brain activation in functional magnetic resonance images. Human Brain Mapping, 7, 1–14.

Everitt, BS and Dunn, G (2001) Applied Multivariate Data Analysis (2nd ed.). London: Arnold.

Everitt, BS, Gourlay, J, and Kendall, RE (1971) An attempt at validation of traditional psychiatric syndromes by cluster analysis. British Journal of Psychiatry, 119, 299–412.

Everitt, BS, Landau, S, and Leese, M (2001) Cluster Analysis (4th ed.). London: Arnold.

Everitt, BS and Rabe-Hesketh, S (1997) The Analysis of Proximity Data. London: Arnold.

Feyerabend, P (1975) Against Method. London: Verso.

Fisher, NI and Switzer, P (1985) Chi-plots for assessing dependence. Biometrika, 72, 253–265.

Fisher, NI and Switzer, P (2001) Graphical assessment of dependence: Is a picture worth 100 tests? The American Statistician, 55, 233–239.

Fisher, RA (1936) The use of multiple measurements on taxonomic problems. Annals of Eugenics, 7, 179–188.

Fraley, C and Raftery, AE (1998) How many clusters? Which cluster method? Answers via model-based cluster analysis. The Computer Journal, 41, 578–588.

Fraley, C and Raftery, AE (1999) MCLUS: Software for the model-based cluster analysis. Journal of Classification, 16, 297–306.

Fraley, C and Raftery, AE (2002) Model-based clustering, discriminant analysis, and density estimation. Journal of the American Statistical Association, 97, 611–631.

Frets, GP (1921) Heredity of head form in man. Genetica, 3, 193–384.

Friedman, HP and Rubin, J (1967) On some invariant criteria for grouping data. Journal of the American Statistical Association, 62, 1159–1178.

Friedman, JH (1989) Regularized discriminant analysis. Journal of the American Statistical Association, 84, 165–175.

Goldberg, KM and Iglewicz, B (1992) Bivariate extensions of the boxplot. Technometrics, 34, 307–320.

Gordon, AD (1999) Classification (2nd ed.). Boca Raton, Florida: Chapman & Hall/CRC Press.

Gower, JC (1966) Some distance properties of latent root and vector methods used in multivariate analysis. Biometrika, 53, 325–338.

Greenacre, M (1992) Correspondence analysis in medical research. Statistical Methods in Medical Research, 1, 97–117.

Hancock, BW, Aitken, M, Martin, JF, Dunsmore, IR, Ross, CM, Carr, I, and Emmanuel, IG (1979) Hodgkin's disease in Sheffield (1971–1976). Clinical Oncology, 5, 283–297.

Hand, DJ (1998) Discriminant analysis, linear. In Encyclopedia of Biostatistics (eds. P Armitage and T Colton). Chichester: Wiley.

Hawkins, DM, Muller, MW, and ten Krooden, JA (1982) Cluster analysis. In Topics in Applied Multivariate Analysis (ed. DM Hawkins). Cambridge: Cambridge University Press.

Heitjan, DF (1997). Ignorability, sufficiency and ancillarity. Journal of the Royal Statistical Society, Series B (Methodological) 59, 375, 381.

Henderson, HV and Velleman, PF (1981) Building multiple regression models interactively. Biometrics, 37, 391–411.

Hendrickson, AE and White, PO (1964) Promax: A quick method for rotation to oblique simple structure. British Journal of Mathematical Statistical Psychology, 17, 65–70.

Heywood, HB (1931) On finite sequences of real numbers. Proceedings of the Royal Society, Series A, 134, 486–501.

Hills, M (1977) Book review. Applied Statistics, 26, 339–340.

Hotelling, H (1933) Analysis of a complex of statistical variables into prinicpal components. Journal of Educational Psychology, 24, 417–441.

Huba, GJ, Wingard, JA, and Bentler, PM (1981) A comparison of two latent variable causal models for adolescent drug use. Journal of Personality and Social Psychology, 40, 180–193.

Hyvarinen, A, Karhunen, J, and Oja, E (2001) Independent Component Analysis. New York: Wiley.

Jennrich, RI and Sampson, PF (1966) Rotation for simple loadings. Psychometrika, 31.

Johnson, RA and Wichern, DW (2003) Applied Multivariate Statistical Analysis: Prentice-Hall.

Jolliffe, IT (1972) Discarding variables in a principal components analysis I: Artificial data. Applied Statistics, 21, 160–173.

Jolliffe, IT (1986) Principal Components Analysis. New York: Springer-Verlag.

Jolliffe, IT (1989) Roation of ill-defined principal components. Applied Statistics, 38, 139–148.

Jolliffe, IT (2002) Principal Component Analysis (2nd ed.). New York: Springer.

Jones, MC and Sibson, R (1987) What is projection pursuit? Journal of the Royal Statistical Society A, 150, 1–36.

Kaiser, HF (1958) The varimax criterion for analytic rotation in factor analysis, Psychometrika, 23, 187–200.

Keyfitz, N and Flieger, W (1971) Population: The Facts and Methods of Demography. San Francisco: W. H. Freeman.

Krause, A and Olsen, M (2002) The Basics of S-PLUS (3rd ed.). New York: Springer.

Krzanowski, WJ (1988) Principles of Multivariate Analysis. Oxford: Oxford University Press.

Krzanowski, WJ (2004) Canonical correlation. In Encyclopedic Companion to Medical Statistics (eds. BS Everitt and C. Palmer). London: Arnold.

Lackey, NR and Sullivan, J (2003) Making Sense of Factor Analysis: The Use of Factor Analysis for Instrument Development in Health Care Research. Sage Publications.

Lawley, DN and Maxwell, AE (1971) Factor Analysis as a Statistical Method (2nd ed.). London: Butterworths.

Little, RJA and Rubin, DB (1987) Statisticial Analysis with Missing Data. New York: Wiley.

Longford, N (1993) Inference about variation in clustered binary data. In Multilevel Conference. Los Angeles.

Mardia, KV, Kent, JT, and Bibby, JM (1979) Multivariate Analysis. London: Academic Press.

Marriott, FHC (1982) Optimization methods of cluster analysis. Biometrika, 69, 417–421.

Mayor, M, Frei, P-Y, and Roukema, B (2003) New Worlds in the Cosmos: The Discovery of Exoplanets. English language edition, Cambridge University Press 2003; originally published as Les Nouveaux Mondes du Cosmos, Editions du Seuil 2001.

McDonald, GC and Schwing, RC (1973) Instabilities of regression estimates relating air pollution to mortality. Technometrics, 15, 463–482.

Morant, GM (1923) A first study of the Tibetan skull. Biometrika, 14, 193–260.

Morrison, DF (1990) Multivariate Statistical Methods (3rd ed.). New York: McGraw-Hill.

Murray, GD and Findlay, JG (1988) Correcting for the bias caused by dropouts in hypertension trials. Statistics in Medicine, 7, 941–946.

Muthen, LK and Muthen, BO (1998) Mplus Users Guide.

O'Sullivan, JB and Mahon, CM (1966) Glucose tolerance test: variability in pregnant and non-pregnant women. American Journal of Clinical Nutrition, 19, 345–351.

Pearson, K (1901) On lines and planes of closest fit to systems of points in space. Philosophical Magazine, 2, 559–572.

Proudfoot, J, Goldberg, D, Mann, A, Everitt, BS, Marks, I, and Gray, J (2003) Computerised, interactive, multimedia cognitive behavioral therapy for anxiety and depression in general practice. Psychological Medicine, 33, 217–227.

Rawlings, JO, Pantula, SG, and Dickey, AD (1998) Applied Regression Analysis. New York: Springer.

Rencher, AC (1995) Methods of Multivariate Analysis. New York: Wiley.

Rohlf, FJ (1970) Adaptive hierarchical clustering schemes. Systematic Zoology, 19, 58–82.

Rousseeuw, PJ (1985) Multivariate estimation with high breakdown point. In Mathematical Statistics and Applications (eds. W Grossman, G Pfilug, I Vincze, and W Wertz). Dordrecht: Reidel.

Rousseeuw, PJ and van Zomeren, B (1990) Unmasking multivariate outliers and leverage points (with discussion). Journal of the American Statistical Association, 85, 633–651.

Rubin, DB (1976) Inference and missing data. Biometrika, 63, 581–592.

Rubin, DB (1987) Multiple Imputation for Nonresponse in Surveys. New York: Wiley.

Schafer, JL (1999) Multiple imputation: A primer. Statistical Methods in Medical Research, 8, 3–15.

Schimert, J, Schafer, JL, Hesterberg, T, Fraley C, and Clarkson, DB (2000) Analysing Data with Missing Values in S-PLUS. Seattle: Insightful Corporation.

Scott, AJ and Symons, MJ (1971) Clustering methods based on likelihood ratio criteria. Biometrics, 37, 387–398.

Sibson, R (1979) Studies in the robustness of multidimensional scaling. Perturbational analyis of classical scaling. Journal of the Royal Statistical Society B, 41, 217–229.

Silverman, BW (1986) Density Estimation for Statistics and Data Analysis. London: Chapman and Hall.

Spearman, C (1904) General intelligence objectively determined and measured. American Journal of Psychology, 15, 201–293.

Spicer, CC, Laurence, GJ, and Southall, DP (1987) Statistical analysis of heart rates and subsequent victims of sudden infant death syndrome. Statistics in Medicine, 6, 159–166.

Tabachnick, BG and Fidell, B (2000) Using Multivariate Statistics (4th ed.). Upper Saddle River, NJ: Allyn & Bacon.

Thurstone, LL (1931) Multiple factor analysis. Psychology Review, 39, 406–427.

Tubb, A, Parker, NJ, and Nickless, G (1980) The analysis of Romano-British pottery by atomic absorption spectroplotomy. Archaeometry, 22, 2, 153–171.

Tufte, ER (1983) The Visual Display of Quantitative Information. Cheshire, CT: Graphics Press.

Velleman, PF and Wilkinson, L (1993) Normal, ordinal, internal and ratio typologies are misleading. The American Statistician, 47, 65–72.

Verbyla, AP, Cullis, BR, Kenward, MG, and Welham, SJ (1999) The analysis of designed experiments and longitudinal data using smoothing splines (with discussion). Applied Statistics, 48, 269–312.

Wand, MP and Jones, CM (1995) Kernel Smoothing. London: Chapman and Hall.

Ward, JH (1963) Hierarchical grouping to optimize an objective function. Journal of the American Statistical Association, 58, 236–244.

Wastell, DG and Gray, R (1987) The numerical approach to classification: a medical application to develop a typoloty of facial pairs. Statistics in Medicine, 6, 137–164.

Wermuth, N (1976) Exploratory analyses of multidimensional contingency tables. In Proceedings 9th International Biometrics Conference, pp. 279–295: Biometrics Society.

Young, G and Householder, AS (1938) Discussion of a set of points in terms of their mutual distances. Psychometrika, 3, 19–22.

Zerbe, GO (1979) Randomization analysis of the completely randomized design extended to growth and response curves. Journal of the American Statistical Association, 74, 215–221.

Index

Springer Texts in Statistics *(continued from previous page)*